DR. S. V. KRUPA
DEPARTMENT OF PLANT PATHOLOGY.
495 BORLAUG HALL
1991 BUFORD CIRCLE
ST. PAUL, MINNESOTA 55108

Sagar V. Krupa, Ronald N. Kickert,
Hans-Jürgen Jäger

Elevated Ultraviolet (UV)-B Radiation and Agriculture

 Springer

Sagar V. Krupa
University of Minnesota
St. Paul, Minnesota, U.S.A.

Ronald N. Kickert
Environmental Consultant
Richland, Washington, U.S.A.

Hans-Jürgen Jäger
Institut für Pflanzenökologie
der Justus-Liebig-Universität
Gießen, Germany

ISBN: 3-540-63892-x Springer-Verlag Berlin Heidelberg New York
Environmental Intelligence Unit

Library of Congress Cataloging-in-Publication Data
Krupa, Sagar V.
    Elevated ultraviolet (UV)-B radiation and agriculture / Sagar V. Krupa, Ron Kickert, H-J Jäger.
    p. cm.—(Environmental intelligence unit)
    Includes bibliographical references and index.
    ISBN 1-57059-475-9 (alk. paper).—ISBN 3-540-63892-x
    1. Crops—Effect of ultraviolet radiation on. 2. Crops and climate. 3. Climate changes. I. Kickert, R.N. II. Jäger, H.J. III. Title. IV. Series.
S600.7.S65K78 1998   97-47088
632'.3—dc21   CIP

This work is subject to copyright. All rights are reserved, whether the whole or part of the material is concerned, specifically the rights of translation, reprinting, reuse of illustrations, recitation, broadcasting, reproduction on microfilm or in any other way, and storage in data banks. Duplication of this publication or parts thereof is permitted only under the provisions of the German Copyright Law of September 9, 1965, in its current version, and permission for use must always be obtained from Springer-Verlag. Violations are liable for prosecution under the German Copyright Law.

© Springer-Verlag Berlin Heidelberg and Landes Bioscience Georgetown, TX, U.S.A. 1998
Printed in Germany

The use of general descriptive names, registered names, trademarks, etc. in this publication does not imply, even in the absence of a specific statement, that such names are exempt from the relevant protective laws and regulations and therefore free for general use.

Product liability: The publisher cannot guarantee the accuracy of any information about dosage and application thereof contained in this book. In every individual case the user must check such information by consulting the relevant literature.

Typesetting: Landes Bioscience Georgetown, TX, U.S.A.

SPIN 10661565   31/3111 - 5 4 3 2 1 0 - Printed on acid-free paper

PREFACE

After a series of scientific hearings, in 1992 the Formal Commission on "Protecting the Earth's Atmosphere" of the Parliament of the Federal Republic of Germany (FRG, Bonn) invited the first two authors to prepare a comprehensive assessment report on the "Effects of Elevated Ultraviolet (UV)-B Radiation on Agricultural Production." During that effort, the third author of this book served as a technical member of the Formal Commission.

The invitation represented a scientific challenge, because any increases in the surface level UV-B radiation across all latitudes are considered to be only one of the many aspects of global change. Although human influences on climate modification are considered to be critical, in turn any observed or predicted climate change is expected to alter human habits, and political and socio-economic patterns in the future. Thus, there is a feedback mechanism within the system of global change. This will reflect on future agricultural patterns.

The emphasis of the report of Krupa and Kickert was on FRG and the countries of the European Economic Community. In comparison, the contents of this book are directed to global scale climate issues, with emphasis on UV-B and world agriculture. Because of the paucity and uncertainty in the literature base, in some cases we were forced to rely on our best judgement. We hope that the reader can appreciate this and examine the contents of this book within that context.

Uncertainty is an accepted principle in biological research. However, it is our hope that future research on the effects of elevated UV-B radiation and global change will utilize improved experimental methods and approaches to better address and/or reduce this uncertainty.

The first two authors are deeply grateful to the Formal Commission of the FRG for financial support in the preparation of the original report. The first author would also like to acknowledge the University of Minnesota Agricultural Experiment Station for its support-in-kind. We thank Leslie Johnson for her meticulous help with word processing and editing and Sidney Simms for his valuable assistance with the illustrations.

Finally, we acknowledge the help of many of our colleagues who freely provided information and advice during the preparation of this book.

CONTENTS

1. **Global Climate and UV-B Radiation** 1
   Introduction 1
   Present Day General Climate Databases 2
   Long-Term Changes in Solar Radiation 2
   Multi-Year Changes in Ambient CO_2 4
   Total Column and Tropospheric O_3 11
   Model Calculations of Total Column
     and Tropospheric O_3 15
   Ground Level UV-B Fluxes 18
   UV-B Measurements 42
   Model UV-B Calculations 45
   Spatial Variation Caused by Clouds, Tropospheric
     Aerosols and O_3 50
   Inaccuracy and Incomplete Processing of UV-B Data 58
   Timing and Duration of UV-B Measurements 59

2. **Assessment Methodology** 67
   Introduction 67
   Instrumentation for UV-B 67
   Methodology for the Exposure of Crops 71
   UV-B Dosimetry 75
   A Comparative Analysis of O_3-CO_2-UV-B Dosimetry 89
   Altered Ambient CO_2, UV-B, O_3, Temperature
     and Moisture Levels 93

3. **Elevated UV-B Radiation and Crops** 105
   Introduction 105
   Primary Effects of Elevated UV-B on Crops 106
   Sensitivity Rankings of Crop Species 108

4. **Mechanisms of Action** 133
   Introduction 133
   Mechanisms of Action of UV-B and Ozone 137
   Eco-Physiological Adaptations and Evolution 141

5. **Crop-Weed Interactions** 147
   Introduction 147
   The Nature of Competition Between Two Plant Species 148
   The Database on Crop-Weed Competition 149
   Major Crops and Their Principal Weeds 150
   Crops and Weeds Under Increased UV-B Radiation
     and/or Climate Warming 152
   Crops: Possible Winners and Losers 153

6. **Pathogen and Pest Incidence on Crops** 167
    Introduction 167
    Elevated UV-B Levels and Incidence of Pathogens 168
    Elevated Ozone Levels and Incidence of Pathogens 171
    Elevated Ozone or UV-B Levels and Insect Pests 174

7. **Integrated View of Environment-Crop Interactions** 181
    Introduction 181
    Potential Combinations of Interactions 182
    Types of Interactions 182
    General Methods of Interaction Analysis 185
    Experimental Results 186
    Crop Growth Prediction Models 211

8. **World Agricultural Production** 219
    Introduction 219
    Current Agricultural Production 224
    Possible Future Changes in Crop Production 225

9. **Plant Population Genetics** 247
    James V. Groth
    Introduction 247
    The Nature of Intense Cropping Systems 248
    Population Genetics of Crops 250
    Robustness of Populations and Communities
        with Genetic Diversity 251
    Plant Breeding as a Mimic to Natural Processes
        of Adaptation 254
    Seeking Sources of Germplasm that Can Tolerate
        the New Environment 254
    Genotype by Environment Interactions 257
    The Predicted Effects of UV-B on Crop Populations 257
    Long-Term Expectations 258
    Short-Term Expectations 259

10. **Genetic Engineering of Crops** 263
    Introduction 263
    Targeting Potential UV-B Damage in Plants 264
    Methods of Crop Protection 265
    Genes Involved in the Repair of DNA Damaged
        by UV-B Irradiation 265
    Plant Transformation 266
    Agrobacterium-Mediated Transfer Systems 267
    Gene Expression in Transformed Plants 268
    The Role of Molecular Biology 268
    A Perspective of the Future 269

Appendix 1. Common and Scientific Names of Crop Plants in the FAO Database 273

Appendix 2. Countries in the FAO Database 279

Appendix 3. Common and Scientific Names of Plant Species (excluding Chapter 8) 283

CONTRIBUTOR

James V. Groth
Department of Plant Pathology
University of Minnesota
St. Paul, Minnesota, U.S.A.
Chapter 9

NOTE TO THE READER

According to the Commission Internationale de l'Éclairage (CIE), ultraviolet (UV)-B is the solar radiation in the bandwidth of 280 to 315 nm. However, many biologists prefer to use the range of 280 to 320 nm. We are not aware of any consistent discontinuties in biological responses to either 315 or 320 nm. Since the subject of the book is agriculture, we have used the range of 280 to 320 nm, and respectively, 200 to 280 and 320 to 400 nm for UV-C and UV-A.

CHAPTER 1

Global Climate and UV-B Radiation

Introduction

The observed and/or predicted changes in the global climate are of utmost international concern. The issue of global climate change should be viewed as a system of atmospheric processes and their products. Among these phenomena are the measured and/or projected increases in surface level ultraviolet (UV)-B (280 to 320 nm wavelength) radiation. While increases in UV-B can lead to a greater incidence of human melanoma, there is also much concern regarding the adverse effects on other forms of terrestrial (and aquatic) life.

During 1991, the German Bundestag[1] published a document on the depletion of stratospheric ozone (O_3), changes in surface UV-B radiation and their potential effects on human health and natural terrestrial and marine ecosystems. However, that report did not include a discussion of the effects of UV-B radiation on agriculture. Similarly, the Scientific Committee on Problems of the Environment (SCOPE)[2] presented a brief, state-of-the-knowledge assessment, combined with an extensive, basic research implementation plan, but did not specifically consider agriculture. Although in recent years a number of scientific reviews have appeared in the literature on the effects of elevated levels of UV-B radiation on terrestrial vegetation (see chapter 3), in the present assessment, as much as published studies will allow, we direct our attention to the aspects of climate change, UV-B and world agriculture.

Elevated Ultraviolet (UV)-B Radiation and Agriculture, by Sagar V. Krupa, Ronald N. Kickert and Hans-Jürgen Jäger.
© 1998 Springer-Verlag and Landes Bioscience.

Present Day General Climate Databases

A computerized global database at the scale of a half degree latitude by a half degree longitude for average present day sunshine, air temperature and precipitation has been produced by the International Institute for Applied Systems Analysis (IIASA).[3] Maps produced from this database are shown in Figure 1.1. Elsewhere there exist one degree and half degree terrestrial ecosystem databases that can be used with the IIASA climate database, to determine apparent climatic limits of regional scale native vegetation.[4-7] But as yet there are no such integrated databases to enable the determination of apparent crop growth limits from correlations with present day climate patterns for specific field-grown agricultural crops.

A climate analysis for western Europe has been performed based on the ground-level weather records for 76 locations in the Netherlands, Belgium, Luxembourg, Germany, France and Switzerland.[8] This study used complex statistical pattern analyses on four climate variables: precipitation, air temperature, potential evapotranspiration and relative humidity on a bi-monthly basis. Eight climate regions for western Europe were identified. A unique result was that the climate of the valleys of the Rhine, Mosel and Meuse Rivers, from Freiburg, Germany in the south, to the area of De Bilt, Netherlands in the north, was distinguished from all of the other climatic areas as having a low relative humidity during spring and summer. A possible inference from this is that without any future precipitation and/or temperature changes, this climatic region could be vulnerable to increased ground-level UV-B radiation, if sufficient stratospheric O_3 depletion was to occur over northern Europe (at levels greater than the offsetting capacity of tropospheric O_3 and aerosols). The only other notable western European climatic region that could also be inferred to be vulnerable might be the coast from Lille, France in the southwest, to Den Helder, Netherlands and northeast to Nordeney, Germany. This might be possible, for this climatic region was found to be distinguished from others by a much greater prominence of potential evapotranspiration (a substitute variable for solar radiation) during winter and spring.

Long-Term Changes in Solar Radiation

Even though scientific concern regarding possible stratospheric O_3 depletion and the corresponding possible increase in the transmission of UV-B radiation to the ground level was present in the

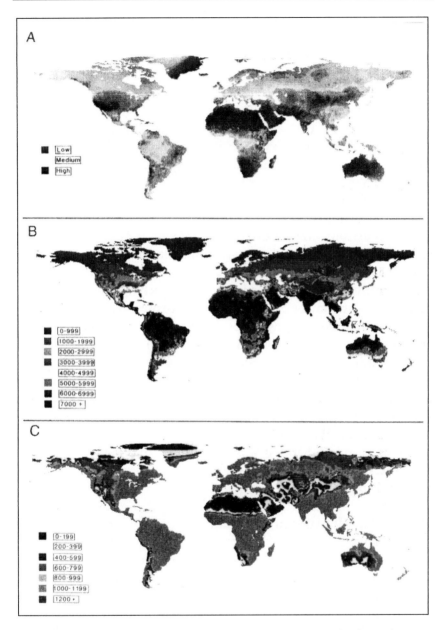

Fig. 1.1. (A) Average annual recorded bright sunshine hours.[3] Note: this figure does not include the level of total radiation or UV-B radiation. (B) Average annual growing degree days (above 5°C).[3] (C) Average annual precipitation (mm).[3] See color insert for color representation.

mid-1970s, there has been no long-term global monitoring network for UV-B radiation.[9,10] However, there has been a long-term network for monitoring "global" solar radiation, The World Radiation Network.[11] In this case, the meteorological term "global" customarily refers to both direct beam and diffuse solar radiation at a given location. The wavelengths of UV-B radiation (280 to 320 nm) are only a small portion (~0.5%) of total global solar radiation arriving at the surface. The World Radiation Network measurements represent the longest data record available and may provide an indication of possible changes in UV-B radiation[12] that may have occurred over the last 27 years (Table 1.1).

According to Stanhill and Moreshet,[11] many of the more than 500 comparisons made between annual values of global radiation measured with thermoelectric pyranometers at the same World Radiation Network stations in 1958, 1965, 1975 and 1985 showed statistically significant reductions in insolation, which were larger than the maximum uncertainty expected in the measurements.

The reductions were greatest in the mid-latitudes of the Northern Hemisphere and greater in summer than in winter and in the first and third periods of comparison. Between 1958 and 1985 an average reduction of 9 W m^{-2} or 5.3% was calculated by two methods as the mean weighted difference in insolation over the earth's land surfaces. This effect may be due to tropospheric aerosol loading, although such a phenomenon is not expected to be uniform across all latitudes. In this context, it should be noted that the data available from the World Radiation Network are very unevenly distributed in time and space.

Multi-Year Changes in Ambient CO_2

Prospects for future climate changes have at least two anthropogenic driving forces: (a) CFCs (chlorofluorocarbons), OBs (organobromines), N_2O (nitrous oxide) and CH_4 (methane) that migrate to the stratosphere and deplete the UV-B absorbing stratospheric O_3; and (b) increases in tropospheric CO_2 and other radiative gases that are anticipated to cause an accelerated "greenhouse effect." It is instructive, therefore, to view the temporal changes in ambient CO_2 (the predominant contributor to the "greenhouse effect") that have taken place over the last several years.[13]

The monitoring locations (Table 1.2) and their corresponding ambient CO_2 concentrations shown in Figures 1.2 through 1.8 are

Table 1.1. Changes in annual insolation between 1958 and 1985 at 46 fixed stations[a,b]

Station	Coordinates		Altitude, m	Population[c] 10^3	K↓[d] 1958, GJ m^{-2} yr^{-2}	Change in %[e]	K↓ MJ m^{-2} yr^{-1}
	Latitude	Longitude					
Resolute A	74°43'N	94°59'W	64	0.2	3.278	−10.5	−348
Wrangle	70°58'N	178°32'W	2	—	3.136	+2.4	+73
Oleněk	68°30'N	112°26'E	127	—	3.403	−8.6	−295
Sodankylä	67°22'N	26°39'E	178	8	2.925	−3.2	−98
Turukhansk	65°47'N	87°57'E	38	—	3.445	+0.3	+15
Arkhangel'sk	64°35'N	40°30'E	8	1402	2.977	+3.0	+92
Reykjavík	64°08'N	21°54'W	66	82	3.129	−1.3	−45
Oimyakon	63°16'N	143°09'E	74	—	4.031	+1.6	+66
Yakutsk	62°05'N	129°45'E	98	108	4.118	−7.3	−300
Jokionen	60°49'N	23°30'E	103	6	3.568	−6.2	−217
Lerwick	60°08'N	1°11'W	82	6	2.807	+5.7	+158
Leningrad	59°58'N	30°18'E	72	3990	3.218	−2.2	−69
Sverdlovsk	56°48'N	60°38'E	290	4310	3.985	−3.0	−115
Omsk	54°56'N	73°24'E	119	1820	4.264	−1.3	−60
Hamburg	53°39'N	10°07'E	49	1793	3.500	−7.0	−247
Potsdam	52°22'N	13°05'E	107	111	3.676	−2.4	−85
De Bilt	52°06'N	5°11'E	46	—	3.431	−0.8	−31
Chita	52°01'N	113°20'E	671	1145	4.879	−8.6	−416
Valentia	51°56'N	10°15'W	14	0.9	3.555	−4.9	−171
Moosonee	51°16'N	80°39'W	10	—	4.017	+1.6	+61
Dresden	51°07'N	13°41'E	271	502	3.507	+2.7	+98
Uccle	50°48'N	4°21'E	105	79	3.444	−0.05	−2
Kiev	50°24'N	30°27'E	121	1630	4.034	+7.3	+289
Khabarovsk	48°31'N	135°10'E	87	437	4.750	−0.8	−43
Wien	48°15'N	16°22'E	202	1870	3.891	+1.3	+53

Location	Latitude	Longitude	Population	KJ	%	
Freiburg	48°00'N	7°51'E	308	4.081	-0.01	-1
Hohenpeissenberg	47°48'N	11°01'E	990	5.028	-14.3	-723
Bol'shaya Elan	46°55'N	142°44'E	22	4.786	-6.2	-295
Locarno	46°10'N	8°47'E	380	4.922	-7.0	-347
Toronto	43°40'N	79°24'W	116	5.121	-6.5	-335
Sapporo	43°03'N	141°20'E	2628	4.897	-7.6	-368
Tateno	36°03'N	140°08'E	117	5.475	-15.7	-866
Bet Dagan	32°00'N	34°54'E	25	7.947	-22.1	-1751
Kagoshima	31°34'N	130°33'E	30	5.525	-12.7	-704
Calcutta	22°39'N	88°27'E	4	6.844	-8.1	-555
Nairobi	1°18'S	36°45'E	10	7.366	+5.4	+403
Bulawayo	20°09'S	28°37'E	1798	7.663	-0.5	-41
Roodeplaat	25°35'S	28°21'E	1344	7.376	+2.4	+189
Pretoria	25°45'S	28°14'E	1164	7.215	+2.4	+166
Bloemfontein	29°10'S	26°11'E	1369	8.295	-5.4	-441
Capetown	33°58'S	18°36'E	1422	7.283	-6.5	-475
Port Elizabeth	33°59'S	25°36'E	49	6.625	-5.9	-393
Wellington	41°17'S	174°46'E	691	5.266	+1.3	+76
Invercargill	46°25'S	168°19'E	387	4.670	-4.6	-213
Mirny	66°33'S	93°01'E	126	5.049	-15.9	-796
South Pole	90°00'S	0°00'	2	4.160	-9.7	-407
			37	All data: Mean	-3.91	-207
			2800	St. deviation	±6.16	±363
				Only increases: Mean	+2.89	+134
				St. deviation	±2.03	±109
				Only decreases: Mean	-6.59	-338
				St. deviation	±5.18	±345

[a]Reprinted with permission from: Stanhill G, Moreshet S. Climate Change 1992; 21:7-75. © 1992 Kluwer Academic Publishers. [b]Locations with increase in annual insolation are highlighted by shaded area. [c]Population data for 1970, data are for metropolitan area. —, represents generally unpopulated areas or small, unlisted settlements. [d]KJ, global irradiance. [e][(1958-1985)/1958] x 100.

Table 1.2. Selected locations across the world of ambient CO_2 measurements

Location	Longitude	Latitude	Characteristic	Elevation (m MSL)
Garmisch-Partenkirchen, Germany	47°28'N	11°03'E	Grassland valley	740
Ascension, U.K. Territory	7°55'S	14°25'W	Island seashore	54
Ragged Point, Barbados	13°10'N	59°26'W	Island seashore	3
Cape Grim, Australia	40°41'S	144°41'E	Promontory seashore	94
Guam, U.S. Territory	13°26'N	144°47'E	Island seashore	2
Niwot Ridge, Colorado, U.S.	40°03'N	140°38'W	Alpine mountain	3749
Seychelle (Mahé Island)	4°40'S	55°10'E	Island seashore	3

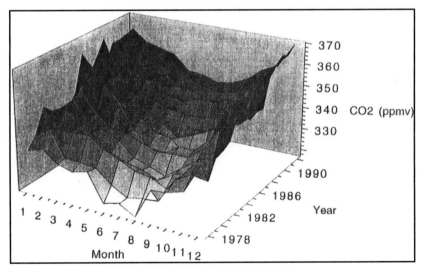

Fig. 1.2. Average monthly ambient CO_2 concentrations at Garmisch-Partenkirchen, Germany (1978-92).[14] Carbon dioxide data were obtained through continuous measurements with a URAS 2T nondispersive infrared (NDIR) analyzer. In 1987, the sampling site was relocated about 0.5 km from the original site. Prior to 1987, measurements were made at a height of 10 m above the surface and after 1987, at 5 m. Monthly values are for daytime measurements only. Data prior to 1987 are not comparable to data following that year. For graphic representation, monthly missing values were replaced using "weighted fill" along months and across years (e.g., 1987).

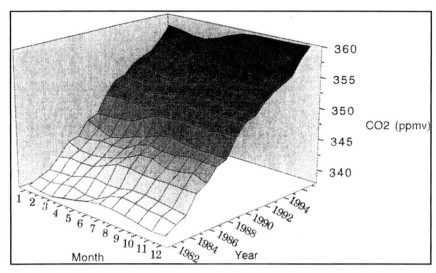

Fig. 1.3. Average monthly ambient CO_2 concentrations at Ascension Island (1980-95), partially from Conway et al.[15] Carbon dioxide mixing ratios were obtained through periodic, intermittent measurements with cylindrical glass flask air samples.

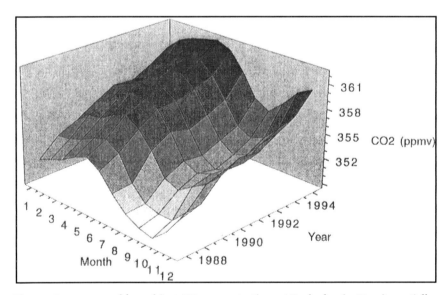

Fig. 1.4. Average monthly ambient CO_2 concentrations at Barbados (1988-95), partially from Conway et al.[15] Carbon dioxide mixing ratios were obtained through periodic, intermittent measurements with cylindrical glass flask air samples.

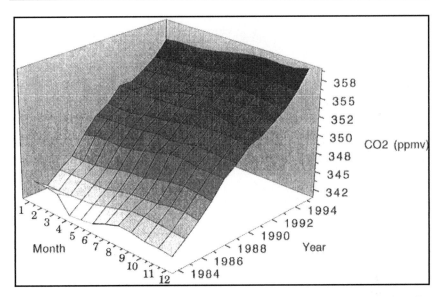

Fig. 1.5. Average monthly ambient CO_2 concentrations at Cape Grim (1984-95), partially from Conway et al.[15] Carbon dioxide mixing ratios were obtained through periodic, intermittent measurements with cylindrical glass flask air samples. For graphic representation, missing monthly data, were replaced using "weighted fill" along months and across years.

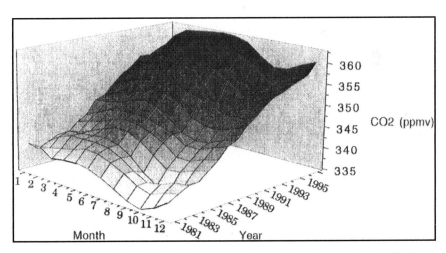

Fig. 1.6. Average monthly ambient CO_2 concentrations at Guam (1979-95), partially from Conway et al.[15] Carbon dioxide mixing ratios were obtained through periodic, intermittent measurements with cylindrical glass flask air samples. For graphic representation, missing monthly data, were replaced using "weighted fill" along months and across years.

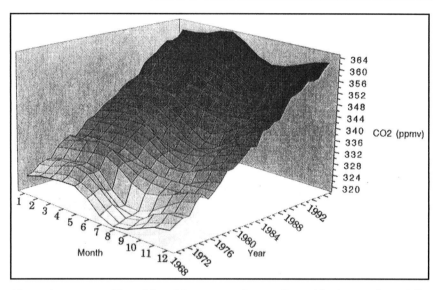

Fig. 1.7. Average monthly ambient CO_2 concentrations at Niwot Ridge (1968-95), partially from Conway et al.[15] Carbon dioxide mixing ratios were obtained through periodic, intermittent measurements with cylindrical glass flask air samples. For graphic representation, missing monthly data, were replaced using "weighted fill" along months and across years.

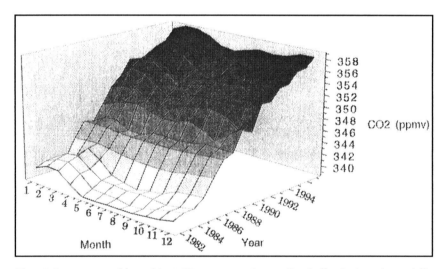

Fig. 1.8. Average monthly ambient CO_2 concentrations at Seychelles (1980-95), partially from Conway et al.[15] Carbon dioxide mixing ratios data were obtained through periodic, intermittent measurements with cylindrical glass flask air samples. For graphic representation, missing monthly data, were replaced using "weighted fill" along months and across years.

actually the net amount left in the atmosphere as a result of a dynamic balance between the sources of CO_2, such as from human activity, and the sinks for CO_2, such as uptake by vegetation. The effect of CO_2 uptake by vegetation can readily be seen across the years at most sites by examining the horizontal axis in the figures.

On an annual basis and as an average across all NOAA (National Oceanic and Atmospheric Administration, U.S.) measurement sites listed in Table 1.2, CO_2 concentrations have been increasing at a rate of 1.43 ppm (~ 0.4%), with a range of 1.10 (Barbados) to 1.43 ppm (Cape Grim). However, at Garmisch-Partenkirchen (site operated by the Fraunhofer Institute, Germany) CO_2 levels have risen at a rate of 1.75 ppm per year since the early 1980s. In a comparative 2-year study of air pollutant concentrations at a background versus a semi-urban monitoring site in Alberta, Canada, the maximum hourly CO_2 concentration at the background site was 381 ppm, while at the second site it was 487 ppm. Thus, although annual average values across sites may be comparable, as with other gaseous pollutants, significant spatial and temporal variabilities should be expected for CO_2 concentrations also.

Total Column and Tropospheric O_3

Information about changes of O_3 concentrations in the stratosphere are important to an understanding of any present and future threats of changes in surface UV-B radiation because O_3 absorbs and decreases UV-B levels at the surface.[16] If total atmospheric column O_3 decreases, with the exception of increasing tropospheric fine particulate aerosols and clouds, one could expect UV-B radiation at ground level to increase. Therefore, to understand what changes might be occurring in the UV-B climate, it is necessary to examine the evidence in the form of measurements of changes in stratospheric O_3 column concentrations.

During the 1990 Southern Hemisphere springtime, at McMurdo Station in Antarctica, total O_3 decreased to 145 Dobson units (DU) and UV radiation increased by 300% of the normal.[17] Total O_3 column data show a statistically significant decrease since 1970 during the winter periods.[18] Summer downward trends are small and not significant. For much of the earth, natural variations from seasonal changes in latitudinal surface level O_3 and fine particle aerosols are still more important controls on UV-B flux to the ground than total O_3 column loss alone.[19] Therefore, it is possible that local variations

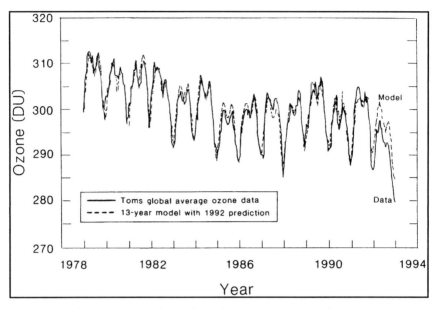

Fig. 1.9. Annual average ozone for the latitude range 65°S to 65°N during 1 January 1979 to 31 December 1992, shown as continuous time series. Each data point is a one week average. The annual solar and quasi-biennial oscillation cycles are clearly evident. In addition to the measured O_3 time series, a statistical model is shown fitted to the 1979 to 1991 time period and extrapolated for 1992. Reprinted with permission from: Gleason JF, Bhartia PK, Herman JR et al. Science 1993; 260:523-526. © 1993 Science.

in air pollution loads and cloud cover have been more important than the losses in stratospheric O_3 for an understanding of UV-B trends since 1970.

Based on the analysis of TOMS (Total Ozone Monitoring Spectrometer) data,[20] latitudes between Seattle and New Orleans, U.S., are losing total O_3 column at the rate of 4 to 5% per decade, with the winter rates being higher, and increased losses lagging during springtime in the Northern Hemisphere[21] (the increase in general radiation amplification factor for DNA-effective UV-B radiation would imply an increase of 8 to 10% per decade). Global total O_3 column appears to be decreasing at a rate of 2.3% per decade.[22]

From satellite measurements with the TOMS on the Nimbus-7 satellite, and Solar Backscatter UltraViolet (SBUV) data, the change in global average total O_3 column over 11 years (October 1978 to November 1989) was found to decrease by 2.9%.[23] The error rate, however, was quite high at ±1.3%. Gleason et al[24] found that the 1992 global average total O_3 was 2 to 3% lower than any earlier observa-

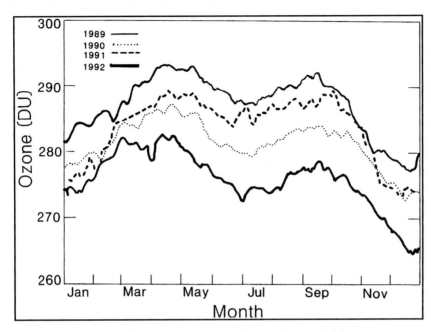

Fig. 1.10. Data on global average O_3 (area-weighted 65°S to 65°N) from NOAA-11-B SUV/2 satellite. Reprinted with permission from: Gleason JF, Bhartia PK, Herman JR et al. Science 1993; 260:523-526. © 1993 Science.

tion by TOMS (1979 to 1991) (Fig. 1.9). According to the authors, O_3 amounts were low in a wide range of latitudes in both the Northern and Southern Hemispheres, and the largest decreases were in the regions from 10°S to 20°S and 10°N to 60°N. These results were confirmed by comparisons with data from other O_3 monitoring instruments: the SBUV/2 on the NOAA-11 satellite, the TOMS on the Russian Meteor-3 satellite, the World Standard Dobson Spectrometer 83 and 22 other ground-based Dobson monitors. In addition, as with the TOMS, the SBUV/2 O_3 data were 2 to 3% lower in 1992 than in 1990, the previous year of lowest O_3 (Fig. 1.10). According to Gleason et al,[24] the cause of the 1992 low O_3 values is uncertain, although the understandable first guess would be that the decrease is related to the continued presence of aerosols from the eruption of Mount Pinatubo.

During January 1992, from the upper atmosphere satellite data, apparently part of the Arctic polar vortex over northern Europe had the highest concentrations of O_3-destroying chlorine monoxide ever found, even compared with the Antarctic O_3 hole.[25] If enough sunlight was present, losses of 30 to 40% of the O_3 could have occurred between January through March. The rule of thumb, or the use of the radiation amplification factor,[18] would have estimated increases of 60 to 80% in DNA-effective UV-B radiation at the ground level over northern Europe at 60°N. Aside from human health concerns, the damage potential of such a phenomenon exists for overwintering crops, such as winter wheat (*Triticum aestivum* L.) and winter barley (*Hordeum vulgare* L.) and permanent pasture land.

For the total column O_3 in the atmosphere, both short-term changes (November 1978 to March 1991) measured by the ground-based Dobson network and TOMS satellite observations and long-term changes (January 1958 to March 1991) measured by the ground-based Dobson network show a loss of UV-B protecting O_3.[20] In the most recent short-term, according to the TOMS data in the Northern Hemisphere, during the summertime the atmosphere over Europe has lost column O_3 at a rate (percent per decade) greater than over North America, the Far East or over the low latitudes. More alarming is the observation that this short-term summertime loss (-4.3% per decade) is more than twice as large as the loss that was found over the long-term period for the same geographic areas (Europe, North America, the Far East, or the low latitudes). Such seasonality of change would certainly be very important for agricultural crops, most of which are either annuals or deciduous perennials. The importance of seasonality with respect to the ability to observe trends has also been the subject of some discussion.[20,26] By the application of the radiation amplification factor at 60°N latitude, the aforementioned loss would translate to an anticipated increase of +8.6% per decade in DNA-effective UV-B at the ground level.

Over Europe during the winter of 1991-92, total O_3 decreased by a record 20%.[27] Similarly, total O_3 data show a statistically significant decrease since 1970 during the winter periods.[28] From the ground-based Dobson spectrometer measurements in Europe, this trend is as large as about -5.3% per decade between 1970 and 1988 at 50°N, but a negative linear trend line would indicate that 55°N generally had a trend of -2.7% ± 1.2% per decade. This is a greater decline than the predictions from the CFC chemistry can explain, so the reasons for this amount of decline are not fully understood. In comparison, summer downward trends are small and not significant.

Model Calculations of Total Column and Tropospheric O_3

Measurements of O_3 in the stratosphere provide us information on what has happened from the recent past to the present. But, by themselves, such measurements cannot provide predictions as to what could happen in the future. For such prognosis, one needs to integrate current knowledge of the physical and chemical processes of the atmosphere, together with the actual measurement data, into a dynamic mathematical model. Such models can be used as tools for experimentation through computer simulation and the development of predictions into the future.

One such model of an idealized 3-dimensional atmospheric computer simulation of the Northern Hemispheric stratosphere, found the formation of an Arctic O_3 hole poleward of 60°N during mid-April springtime. This was similar to that already observed over Antarctica.[29] The O_3 hole was found to occur during exceptionally cold winters, under the current levels of atmospheric chlorine, but with an atmospheric CO_2 concentration that is approximately double (660 ppm) of the current levels. However, it is not clear whether intermediate, lower CO_2 concentrations between 330 and 660 ppm would show the same results. According to Austin et al,[29] under current conditions of chlorine and CO_2, an Arctic O_3 hole does not appear likely. But, since both chemicals are expected to increase in the atmosphere over the coming decades, the study concluded that it is likely that the formation of an O_3 hole over the Arctic region, reaching southward to 60°N is likely, at least during two winters out of every 10 during the next 50 years. Others agree that there is a threat of an Arctic O_3 hole developing by the year 2050.[30]

A recent study used satellite measurements of total column O_3 to modify a standard atmospheric distribution of the O_3 concentration.[31] It was not strictly an observation-based, nor a model-based approach, but a combination of the two. The results showed a decline during the period 1979 through 1989, of slightly over 4% per decade in annual total O_3 at the latitudes of > 35°N (Fig. 1.11). Annual results are not too meaningful because many crops are grown during less than a full year. A different study,[32] used a photochemical model for a longer time period (21 years), from 1966 through 1986, to display the percent seasonal change in total O_3 in the atmosphere (Fig. 1.12). This study showed a 21 year increase in total O_3 during summer and autumn over latitudes of Germany and the other EEC

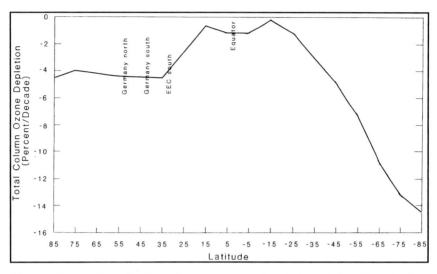

Fig. 1.11. Changes in total column O_3 as percent per decade (1979-89) and by latitude.[31]

countries. However, when the data surfaces in Figure 1.12 are viewed across all of the seasons at the latitudes of Germany and the other EEC countries, an annual decline of -4% in total atmospheric O_3 does appear to match the patterns in Figure 1.11, even though the years of the two studies do not exactly match. The actual reasons for these differences are not known, but they might have resulted from differences in the methods used to process the data between the two studies.

A multiple linear regression, statistical model was parameterized based on 13.2 years of total atmospheric O_3 data from the Nimbus 7 satellite-based sensor.[33] The effect of a multi-year solar cycle on global total atmospheric O_3 was examined and as the solar minimum is approached, the model predicted that more rapid O_3 depletion will occur globally.

If increases occur in the surface UV-B radiation in the future, surely such increases will not be constant, but will fluctuate with the overall weather during the growing season. In between UV-B episodes, field crops will also be exposed to relatively high tropospheric O_3.[22] The evidence shows that over the last 100 years (1880-1980), at nonurban locations in Europe, the monthly mean background surface O_3 has increased by 100% in the winter (from 10 to 20 ppb) and

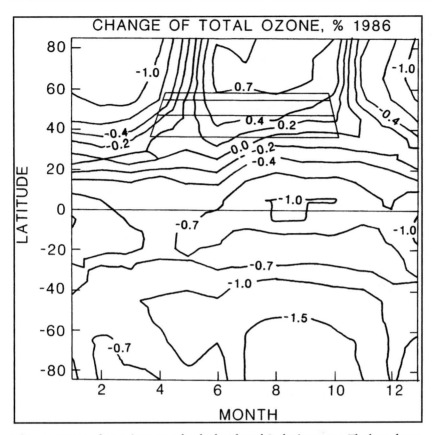

Fig. 1.12. Percent change in seasonal, calculated total O_3 during 1986.[32] The boxed area within the upper portion of the figure represents the latitudes of Germany and the European Economic Community.

by 200 to 350% in the summertime (from 10 to 45 ppb).[34] From 1974 to 1989, Barrow, Alaska, U.S. (71.3°N) has seen an increase of 0.91% ±0.35% per year in monthly mean surface O_3. Projected changes in tropospheric O_3 for the future, based on atmospheric chemistry models vary considerably depending on the assumptions of precursor conditions. Overall, the range of change predicted from various models for tropospheric O_3 in nonurban areas is an increase from about 0.3 to as much as 1% per year during the next 50 years. Compounded annually over the entire 50 years, it would imply increases of from 16 to 64% above the current levels. In comparison, episodal ambient O_3 concentrations would have considerably higher values.

Ground Level UV-B Fluxes

Even though some data show that total atmospheric O_3 has been decreasing, UV-B fluxes at ground level at many locations do not display increasing trends over several years.[10,35,36] A major question is: "What levels of UV-B radiation are being received at ground level in various locations today, and in the last few years?"

The scientific evidence for ground-level UV-B intensities from observation and measurement campaigns, and from computer modeling studies are presented in Tables 1.3 and 1.4. Some of the results pertain to UV radiation at wavelengths outside the commonly accepted 280 to 320 nm. This is indicated in the tables. Some of the data are published without comparison to the wavelength-based variability in the sensitivity of various biological receptors, namely without being weighted by an action spectrum to obtain the effective exposure for some biological process. This is also indicated in the tables. Nonetheless, some of the data in Tables 1.3 and 1.4 should be helpful in evaluating whether various crop species appear to be at risk for having an exposure threshold(s) exceeded.

The data also help to define the UV and UV-B climates around the world. In Table 1.3, the data are from observations, and in Table 1.4, the data are from computer model simulations. The UV detection technique is given for the observed data in Table 1.3. Differing ways for expressing and summarizing UV-B records are used among the many publications. These are listed in the tables as "UV-B Statistic-Definition." Often, summarizing the information of interest to a radiation meteorologist is not very useful to an agricultural crop ecologist. The time-dimensions to which the data refer, are also listed in Tables 1.3 and 1.4. Finally, some of the salient features of the UV-B results are listed along with their sources of reference.

According to Madronich,[31] in the absence of direct measurements of global trends in UV-B, the best estimates are probably obtained by combining the measured ozone trends with radiative transfer calculations (Fig. 1.13). Mid-latitude increases in UV-B occur at all times of the year with a range of 5 to 20% per decade (Table 1.5). The effect of the breakup of the polar vortex is reflected by the increased UV-B levels at latitudes between 30°S and 60°S during December-January, the period of highest natural irradiation. No significant trends are seen in the tropics, since tropical UV levels are naturally high.

Table 1.3. Published observations of ground-level UV and UV-B (ambient exposure values)

Location(s)	Detection technique	Actual irradiance or weighted effective exposure	UV statistic-definition	Temporal dimensions	Numerical results	Reference
European Economic Community						
Neuherberg, Germany (48.2°N, 11.5°E) Lauder, New Zealand (45°S, 170°E)	Double monochromator with spectral resolution of 1 nm	Data for actual spectral irradiance only (W m^{-2} nm^{-1})	Irradiance by wavelength at 34.3° solar zenith angle	1991; measured UV-B only shown for 21 February at Lauder, and July 29 at Neuherberg	Summertime clear sky DNA effective UV-B at Lauder was 1.9 x Neuherberg; Lauder level is from stratospheric O$_3$ decline; Neuherberg level is due to increase in tropospheric O$_3$	Seckmeyer and McKenzie[37]
Reading, southeast England (51.28°N, 0.59°W)	Optronics 742 spectro-radiometer	Actual	280-320 nm; W m^{-2}	19 July 1989 to 31 October 1990; noon, clear days only	Broadband maxima around 3.25 W m^{-2} at noon on clear days for July-August 1989, and 2.25 W m^{-2} at noon during clear days in May-July 1990; will be many years before conclusions about UV-B changes can be made at Reading	Webb[38]

Table 1.3 continued

Location(s)	Detection technique	Actual irradiance or weighted effective exposure	UV statistic-definition	Temporal dimensions	Numerical results	Reference
Reading, southeast England (51.28°N, 0.59°W)	Optronics 742 spectro-radiometer	Actual, 280-320 nm	Irradiance at 300 nm, and total broad-band UV-B	23 August and 29 November 1989, clear sky	Hourly peak about 2.2 W m^{-2} at 1100 h on 23 August 1989; hourly peak about 0.5 W m^{-2} at 1100 h on 29 November 1989	Webb[39,40]
Reading, southeast England (51.28°N, 0.59°W)	Optronics 742 spectro-radiometer	Actual	UV-B irradiance by wavelength	11 July, 28 September, 29 November (1989)	Strong relation of UV-B (320 nm) with total UV-B exposure; from graph: July, September, November daily peak irradiance at 320 nm = 0.275 W m^{-2} nm^{-1} at 1100 h and at 300 nm = 6.5 * 10E^{-3} nm^{-1} at 1200 h; data imply broadband daily peak UV-B values of 2.06, 1.16, and 0.19 W m^{-2}	Webb[41]
Helsinki (60.8°N, 23.5°E), Sodankylä (67.4°N, 26.6°E), Finland	Solar Light Model 501 Biometer	Erythemal weighted spectrum (clear days)	Broadband daily totals	April-May 1993, 1994	Average increase of 10%, max increase 34%	Jokela et al[42]

Location	Instrument	Measurement	Period	Results	Reference
Athens, Greece (38°N, 23.44°E)	Eppley pyranometer; Bird and Riordan[43] and Green[44] models, with Leckner[45] for aerosol transmission function	Actual irradiance (290-380 nm) and calculated (280-400 nm) UV-B and UV-A	15 cloudless days between 21 May 1989 to 14 February 1990	Total integrated UV irradiance by hour (W m^{-2}) From diurnal graphics - interpolated daily peak UV irradiances (W m^{-2}) observed (o), calculated (c): 42 (o),41(c) on 21/5; 44(o),43(c) on 31/5; 45(o)(c) on 7/6; 42(o), 41(c) on 23/6; 42(o),41 (c) on 29/6; 42(o)(c) on 9/7 and 11/7; 43(o), 44(c) on 5/8; 43(o),44(c) on 16/8; 37(o),40(c) on 21/8; 43(o),44(c) on 5/9; 33(o), 37(c) on 20/9; 24(o), 25(c) on 23/10; 17.2(o), 17.8(c) on 1/12; 19(o), 24(c) on 14/2; no statistical results	Katsambas et al[46]
Thessaloniki, Greece (40.5°N, 23°E)	Brewer (single or double monochromator) spectrophotometer	Erythemal weighted and 305 nm (63° SZA)	1991-95 Monthly means	4% per year increase at 305 nm	Zerefos et al[47]
Other European Countries					
Longyearbyen, Svalbard, Norway (77°N, 20°E)	Not given	Actual 290-315 nm; wavelength-integrated diurnal irradiance	11 July - 3 August 1989	Maximum irradiance 0.6 W m^{-2} at midday on 25 July, but otherwise varies down to about 0.14 W m^{-2} daily maxima	Henriksen et al[48]

Table 1.3 continued

Location(s)	Detection technique	Actual irradiance or weighted effective exposure	UV statistic-definition	Temporal dimensions	Numerical results	Reference
Longyearbyen, Svalbard, Norway (77°N, 20°E)	Double monochromator	Actual	290-315 nm; broadband integrated by hour and day	23 July - 4 August 1987	Peak irradiance of 0.230 W m^{-2} on 25 July at noon; 13 day average peak 0.168 W m^{-2} at 1300 hours	Henriksen et al[49]
Longyearbyen, Svalbard, Norway (77°N, 20°E); Tromsø, Norway (69.6°N, 19.2°E)	Double monochromator	Actual	Instantaneous irradiance at solar elevations of 20° (Tromsø) and 30° (Longyearbyen)	July-August 1987	0.120 W m^{-2} (Tromsø), 0.240 W m^{-2} (Longyearbyen)	Stamnes et al[50]
Jungfraujoch, Switzerland (47.31°N, 7.98°E, 3586 m)	Eppley spectrometer with 8 interference filters	Actual	Spectral irradiance at 305-325 nm by 5 nm	Ratios of 315 nm/305 nm for 1985-88, but absolute spectral values only for 21 August 1985	Approximate daily peak irradiances at true solar noon for 21 August 1985: 0.05, 0.15, 0.21, 0.34, and 0.41 W m^{-2} nm^{-1}, at 305, 310, 315, 320, and 325 nm respectively	Blumthaler and Ambach[51]

Location	Instrument	Spectral range	Period	Observations	Reference
Jungfraujoch, Switzerland (46.6°N, 7.98°E, 3586 m); Fairbanks, Alaska (64.8°N, 147.9°W, 133 m)	Eppley UV radiation detectors; 295-385 nm (UV-B and UV-A)	Actual	Daily mean broadband (295-385 nm) values	10 years, 1980 (J) and 1979 (F) to present	Peak daily means annually approximate 21.5 (June) and 12.5 W m^{-2} (July) for Jungfraujoch and Fairbanks, respectively	Ambach et al[52]
Jungfraujoch, Switzerland (47°N, 7.98°E, 3576 m)	Robertson-Berger meter	Erythemal weighted, 295-330 nm	Daily totals of flux ratio of UV-B to total solar radiation (sunburn units, MJ m^{-2})	13 periods, about 8 weeks each, cloudless days over 1981-89	Assuming no decrease in total solar radiation between 1981-89 (but, see Stanhill and Moreshet[11]), at the 1981-89 maximum ratio (0.8), erythemal weighted UV-B is assumed to have increased by 0.7% ±0.2% per year; at the 1981-89 mean of the ratio, the UV-B increase would be 1.1 ±0.4% per year	Blumthaler and Ambach[53]

Table 1.3 continued

Location(s)	Detection technique	Actual irradiance or weighted effective exposure	UV statistic-definition	Temporal dimensions	Numerical results	Reference
Atlantic Ocean cruise along 30°W, from about 37°N to 30°S, south of the Azores, off the west coast of Africa, and southwest of the EEC	Noncommercial instrument described in Dehne[54]	Erythemal weighted, 295-330 nm UV-B	Daily sums (Wh m^{-2}) along 30°W by latitude; ratio of erythemal UV-B to total solar radiation	Per minute (daytime), 15 days 21 September–5 October 1988	Maximum daily sum of 1.45 Wh m^{-2} at the equator; 1.0 to about 0.6 Wh m^{-2} at 30°N to 37°N, southwest of EEC; maximum is in approximate agreement with theoretical computations	Dehne,[54] Behr[55]
Middle East Countries						
Dead Sea shore (–400 m BSL) and Be'er Sheva', Israel (250 m ASL)	Double monochromator spectroradiometer	Actual	Ratio of intensities of lower to higher altitude location by wavelength (300-320 nm); no absolute values	About 11:00 a.m., 1:00 p.m., and 2:30 p.m. February 1987	Higher elevation location has higher intensity in morning and early afternoon; lower elevation location has higher intensity in mid-afternoon	Kushelevsky[56]

| Baghdad, Iraq (33.42°N, 44.42°E) | Eppley UV radiometer (295-385 nm; combined UV-B 295-320 nm, and UV-A 320-385 nm) | Actual | Total UV-B and UV-A hourly fluxes (Wh m^{-2}) by daylight hour; mean, maximum, and minimum daily sums hourly irradiance values by month | Hourly UV values for 22 August and 1 October 1988, and 18 January and 15 April 1989; daily totals by month from August 1988 to April 1989 | Highest hourly intensity of 41.9 Wh m^{-2} on 22 August 1988; lowest hourly intensity, 17 Wh m^{-2}, on 18 January 1989; at the 1988 winter solstice, during a totally cloudy day, daily total UV radiation decreased 63% compared to a clear day 4 days later; maximum total daily UV radiation by month occurred in August (315 Wh m^{-2}), and minimum under cloudy rainy conditions in December (15 Wh m^{-2}); monthly means of total daily UV radiation varied from 80 Wh m^{-2} (December) to 297 Wh m^{-2} (August); the highest monthly proportion UV to total solar was 4.8% (August) | Akrawi and Ahmed[57] |

Table 1.3 continued

Location(s)	Detection technique	Actual irradiance or weighted effective exposure	UV statistic-definition	Temporal dimensions	Numerical results	Reference
Safat, Kuwait (29.5°N, 48°E)	Polysulfone film	Weighted to film sensitivity, roughly erythemal and correlated to spectro-radiometer data for 280-320 nm	11:30 a.m.-noon, film sensitive UV-B irradiance	Half-hour every other day, August 1983 to November 1985	Peak half-hour exposure for time period about 1.56 Wh m^{-2} in May 1985; annual 1985 total UV-B irradiance 1.42 x 10^3 J cm^{-2}	Kollias et al[58]
Canada Toronto, Ontario (44°N, 79°W)	Brewer spectro-photo-meter (model 14)	290-325 nm, with a resolution of 0.5 nm	Daily flux at 300 and 324 nm	1989-93	35% increase per year in the winter and 7% increase per year in the summer at 300 nm	Kerr and McElroy[59]

Location	Instrument	Method	Parameters	Period	Results	Reference
United States of America						
Bethesda, MD (39°N, 77.05°W)	Goldberg custom-made precision spectral UV radiometer, Correll et al[60]	Special elaborate data signal conditioning applied between data collection and final analysis	Maximum monthly mean daily total UV-B flux (J m^{-2} day^{-1}) (295-320 nm); maximum monthly mean solar noon UV-B flux (J m^{-2} min^{-1}) (295-320 nm)	15 year record, 1976 - 1990	Maximum monthly mean daily total 72,490 J m^{-2} day^{-1} in 1986; maximum monthly mean solar noon flux 188 J m^{-2} min^{-1} in 1986	Correll et al[60]
Seguin, TX (29.36°N, 97.55°W)	Total O$_3$ portable spectrometer	300 nm, 306 nm (cloudiness: +6.9; +3; 18.1 and +13.4%; May-August)	30 day running mean	1990-94	Mean direct irradiance at 300 nm was 18.1 and 13.4% higher in July and August 1993 than during the same months in 1990	Mims et al[61]

Table 1.3 continued

Location(s)	Detection technique	Actual irradiance or weighted effective exposure	UV statistic- definition	Temporal dimensions	Numerical results	Reference
Usually airports at Tallahassee, FL (30.26°N, 84.19°W), El Paso, TX (31.84°N, 106.5°W), Fort Worth, TX (32.45°N, 97.2°W), Albuquerque, NM (35.05°N, 106.38°W), Oakland, CA (37.5°N, 122.15°W), Philadelphia, PA (40°N, 75.10°W), Minneapolis, MN (45°N, 93.15°W), Bismarck, ND (46.5°N, 100.48°W)	Robertson-Berger meter	Erythemal weighted (290-330 nm)	Annual sum of R-B counts per 10,000	1974 through 1985	All locations show declines in annual values over the 12 years of measurements; of the 8 locations, the highest annual values were found for El Paso, beginning at about 223 (R-B counts per 10,000) in 1974 and decreasing to 198 in 1985; largest average trend was -1.1% per year for Minneapolis, MN	Scotto et al[36]

Global Climate and UV-B Radiation

Location	Instrument	Measurement	Date/Time	Results	Reference
Rockville, MD (39.05°N, 77.1°W), Hampton, VA (37.02°N, 76.23°W), Barrow, AK (71.16°N, 156.5°W), and Panama (8.57°N, 179.3°W)	Two ground-based Eppley pyranometers and cutoff filters at 295 nm and 400 nm (UV-B and UV-A) with 3 min time interval (Rockville only)	Actual UV-B and UV-A W m^{-2} every 3 min for Rockville; other stations are average daily total UV-B (KJ m^{-2}) by month at 290-325 nm	16 year record for Rockville, at 3 min intervals; other stations are shown monthly for an unspecified year	16 year Rockville record shows 11 year solar cycle with maximum of about 48 W m^{-2} for 3 min in 1972, and about 34 W m^{-2} in 1982; monthly average daily total maxima are about: 90 KJ m^{-2} day^{-1} in June for Panama, 57 KJ m^{-2} day^{-1} in April and June for Rockville, 50 KJ m^{-2} day^{-1} in July for Hampton, and 47 KJ m^{-2} day^{-1} in May and June for Barrow	Goldberg[62]
Mauna Loa, HI (19.5°N), El Paso, TX (31.8°N), Philadelphia, PA (40.0°N), Bismarck, ND (46.8°N)	Robertson-Berger meter	Erythemal weighted (290-330 nm) Approximate peak erythemal irradiance (W m^{-2})	1974, local noon	0.38 W m^{-2}, April, Mauna Loa; 0.31 W m^{-2}, July, El Paso; 0.25 W m^{-2}, July, Philadelphia; 0.21 W m^{-2}, July, Bismarck	Frederick and Snell[63]

Table 1.3 continued

Location(s)	Detection technique	Actual irradiance or weighted effective exposure	UV statistic-definition	Temporal dimensions	Numerical results	Reference
Ephrata, WA (47.2°N), Miami, FL (25.8°N), La Selva, Costa Rica (10.4°N), Irazú, Costa Rica (10°N, 3400 m)	LI-1800 spectro-radiometer	Actual (300-320 nm)	Mean irradiance at solar maximum (MISM) (all locations); Miami June monthly average; Miami total daily mean (TDM), maximum (TDX), range (TDR) for summer and winter	Ephrata, 5 days, June 1986; Miami, 1986-87 winter, 36 days, 1986-87 summer, 39 days; La Selva, April 1983/84/86, 6 days; Irazú, April 1986, 1 day	Ephrata MISM 3.2 W m^{-2}; Miami MISM summer 3.7 W m^{-2}, winter 1.8 W m^{-2}; La Selva MISM 4.3 W m^{-2}; Irazú MISM 6.2 W m^{-2}; Miami June average 4 W m^{-2}; summer TDM 1.8 W m^{-2}, TDX 2.67 W m^{-2}, TDR 1.4-2.47 W m^{-2}; winter TDM 0.8 W m^{-2}, TDX 1.1 W m^{-2}, TDR 0.3-0.9 W m^{-2}	Lee and Downum[64]

Southern Hemisphere

Location	Instrument	Measurement	Dates	Results	Reference
Melbourne, Australia (37.45°S, 144.58°E)	Broadband detectors of Australian Radiation Laboratory	Actual (data given), and "effective irradiance" (no data given); integrated broadband 285-315 nm	10 December 1987 to 19 January 1988	During 10-14 December, when 10.5% decrease in total O_3 occurred, 11.5% increase in UV-B irradiance, and 21% increase in effective irradiance; ozone remained low until 19 January; clear day solar noon irradiance maximum about 2650 mW m^{-2}; total 41 day UV-B irradiance about the same as in previous year because of more cloudiness in 1987-88	Roy et al[65]
Palmer Station, Antarctica (64.46°S, 64.05°W)	Double monochromator, scanning spectroradiometer	DNA-effective daily combined UV-B and UV-A exposure (280-400 nm)	Mid-September-December 1988; mid-August-November 1989; mid-August-December 1990; hourly measurements	Approximate graphic interpolation: absolute seasonal DNA-effective daily exposures (J m^{-2}): 35 observed vs. 24 normal, day 314, 1988; 53 vs. 24 normal, day 302, 1989; 130 vs. 50 normal, day 335, 1990; minimum local noon O_3 (DU) 201 (1988), 170 (1989), 172 (1990); days DNA daily exposure > daily maximum with no O_3 depletion: 36 (1988), 44 (1989), 90 (1990); days DNA exposure > max. normal summer solstice (day 356): 0 (1988), 1 (1989), 20 (1990)	Lubin et al[66]

Table 1.3 continued

Location(s)	Detection technique	Actual irradiance or weighted effective exposure	UV statistic-definition	Temporal dimensions	Numerical results	Reference
Auckland (37°S, 175°E), Wellington (41°S, 175°E), Christchurch (43.5°S, 173°E), New Zealand	International Light Inc. type SED 240/ ACTS 270/W radiometers	Erythemally weighted irradiance at 310 nm (SIA 30° and 45°)	Annual mean irradiance; 5 day maxima	1988-95	Spring to autumn increase in each year, no monotonic trend between years	Ryan et al[67]

Table 1.4. Published model calculations of ground-level UV and UV-B exposure values

Location(s)	Calculation technique	Actual or weighted effective exposure	UV statistic definition	Temporal dimensions	Numerical results	Reference
Hohenpeissenberg, Bavaria, Germany (47°N) (HG), and global by latitude (GL)	Delta-two-stream spectral diffuse radiative transfer model with spherical UV irradiance and two-dimensional photochemical model	Broadband (290-320 nm) UV-B irradiance, DNA-effective exposure, and Robertson-Berger-effective UV-B irradiance	At HG: noon, and daily total exposure, UV-B irradiance and DNA-effective exposure; for GL: 1966 daily UV-B noon irradiance (W m^{-2}); 1966-86, % change in daily total UV-B irradiance; 1966-86, % change in daily total DNA-effective exposure; 1966-86, % change in daily total Robertson-Berger-effective UV-B irradiance	At HG on June summer solstice in 1967/68/78/82; for GL: 1966 noon irradiance, % change in daily total over 1966-86	At HG, highest (H) and lowest (L) values for summer solstice, actual (A) and DNA-effective (D), at noon (N) and daily total (T): 3.2551 (H,A,N) (W m^{-2}), 3.1419 (L,A,N); 2.9128 (H,D,N) (cW m^{-2}), 2.6296 (L,D,N); 24.121 (H,A,T) (Wh m^{-2}), 23.279 (L,A,T); 17.748 (H,D,T) (cWh m^{-2}), 16.069 (L,D,T); for GL by visual interpolation for Germany latitudes for months 4 to 9, 1966 daily UV-B noon irradiance range 1.6 to 2.2 W m^{-2}; 1966-86, % change in daily total UV-B irradiance range -0.5% to -1.0%; 1966-86, % change in daily total DNA-effective exposure range -1.0% to -2.0+%	Brühl and Crutzen[32]

Table 1.4. continued

Location(s)	Calculation technique	Actual or weighted effective exposure	UV statistic-definition	Temporal dimensions	Numerical results	Reference
Global	Multi-layer delta-Eddington radiative transfer scheme,[68] and fast tri-diagonal matrix solution;[69] WMO[70]	Erythemal, plant damage (UV-B), DNA, and Robertson-Berger action spectra for respective portions of 280-400 nm wavelengths (UV-B and UV-A)	Cloudless, aerosol-free sky, change per decade in cumulative annual integration of absolute daily exposures by latitude multiple of 5; % change per decade in cumulative annual integration of various weighted effective daily exposures by latitude multiple of 5	11 years, 1979-89; annual; monthly only for DNA effective integrated exposure	See figures accompanying this text; statistically significant increase of DNA effective daily exposure (J m^{-2} decade^{-1}) between 30°N-60°N latitude band in late winter and early spring, and north of 50°N in summer; largest Northern Hemisphere (NH) increase is 250 J m^{-2} decade^{-1} around 30°N in February; largest statistically significant relative increase in DNA effective daily exposure in NH is 20%+ per decade from 30°N-50°N in December-January; % change per decade annual total plant effective UV-B exposure in NH between 35°N-55°N is 7.0% (±2.7) to 8.1% (±3.0), without troposphere ozone pollution and S-aerosols, these values could have been 4% higher; values are much higher than	Madronich[31]

Global Climate and UV-B Radiation

Latitude bands: 30°N–39°N, 40°N–52°N, 53°N–64°N; European sites: Belsk, Poland; Uccle, Belgium; Hohenpeissenberg, Germany; Arosa, Switzerland	Frederick and Lubin[71] radiative transfer model	UV-B and UV-A over 290–400 nm weighted by erythemal and mouse carcinoma (MC) action spectra; MC spectra results given here because these action spectra are closer to plant action spectra	Annually integrated effective UV irradiance under cloudless pollution-free sky		previously thought; radiation amplification factor for annual integrated plant effective UV-B is 1.60 at 45°N (equal to DNA RAF) and 1.84 at 55°N (greater than DNA RAF)	
			Monthly average, annual average, and annual residuals over 32 years, 1957-88	Approximate average July peak MC effective UV from low-high latitude bands: 0.83, 0.70, 0.50 Wh m^{-2}; 32 year average annual total MC effective UV from low-high latitude bands: 1.79, 1.12, 0.65 Wh m^{-2}; no statistically significant trends at any latitude bands over the total 32 years; for shorter periods, the only significant trend was at 40°–52°N for 1970-88 where annual integrated MC effective UV increased 3.1% ±1.8% per decade; latitudes of southern Germany and central EEC; at European sites, only Uccle and Arosa show 1970-88 statistically significant trends in annual total MC effective UV of +5.06 ±2.86% and +4.37 ±2.80% per decade	Frederick et al[72]

Table 1.4. continued

Location(s)	Calculation technique	Actual or weighted effective exposure	UV statistic-definition	Temporal dimensions	Numerical results	Reference
Latitude bands: 35°N, 45°N, and 55°N; (Germany lies within 45°N-55°N, the EEC countries lie within 35°N-55°N)	Applies January and July 1970 total atmosphere O₃ data, and the trends described in Frederick,[28] to the year 1990 in a radiative transfer model	UV-B flux (W m^{-2}) weighted by UV-induced mouse carcinoma action spectra	Mouse-carcinoma effective irradiance	Clear sky, local noon, January and July	Latitude and season are the determinants of effective UV-B irradiance between years 1970-90 at 35°N to 55°N; the "changes" in effective UV-B irradiance shown by ozone depletion across these latitudes over this period are insignificantly small	Frederick[28]
Global by latitude	Multi-layer atmospheric radiative transfer model	Actual clear sky UV-B irradiance (290-320 nm)	Instantaneous irradiance at 10:00 a.m.	Daily over a year	Graph interpolated spring, summer and fall values for Germany: 2.0 to 3.2 to 1.0 W m^{-2}; for EEC south to north by season: 3.0 to 2.0, 3.9 to 2.6, 2.0 to 0.8 W m^{-2}	Serafino and Frederick[73]

Location	Method	Measurement	Period	Results	Reference
Oslo, Norway, 60°N; latitude bands 30°N-39°N, 40°N-52°N, and 53°N-64°N	Discrete ordinate radiative transfer model[74]	International Electrotechnical Commission action spectra, similar to erythemal; 290-400 nm (UV-A and UV-B)	Relative annual effective dose; % change in annual UV dose	1978-88 (Oslo); 1969-86	No trend in effective UV exposure over 1978-88 at Oslo; % change in annual effective UV exposures from 1969-86 for latitude bands 30°N-39°N, 40°N-52°N, and 53°N-64°N: +2.0%, +3.1%, +0.5%, respectively	Dahlback et al[75]
Athens, Greece, 37.97°N	Eppley pyrano-meter; Bird and Riordan[43] and Green[44] models, with Leckner[45] for aerosol transmission function	Actual irradiance (290-380 nm) and calculated (280-400 nm) UV-B and UV-A	Total integrated UV irradiance by hour (W m^{-2})	15 cloudless days between 21/5/89 to 14/2/90	From diurnal graphics - interpolated daily peak UV irradiances (W m^{-2}) observed (o), calculated (c): 42(o),41(c) on 21/5; 44(o), 43(c) on 31/5; 45(o)(c) on 7/6; 42(o), 41(c) on 23/6; 42(o), 41(c) on 29/6; 42(o)(c) on 9/7 and 11/7; 43(o), 44(c) on 5/8; 43(o), 44(c) on 16/8; 37(o), 40(c) on 21/8; 43(o),44(c) on 5/9; 33(o), 37(c) on 20/9; 24(o), 25(c) on 23/10; 17.2(o),17.8(c) on 1/12; 19(o), 24(c) on 14/2; no statistical results	Katsambas et al[46]

Table 1.4. continued

Location(s)	Calculation technique	Actual or weighted effective exposure	UV statistic-definition	Temporal dimensions	Numerical results	Reference
Nonurban Europe and eastern United States	Discrete ordinate radiative transfer model[74]	DNA action spectrum (280-400 nm; UV-B and UV-A)	Proportion reduction in integrated daily DNA effective exposure	Instantaneous 280-400 nm exposures at 15 min time steps; time span covers total time since pre-industrial era, no decadal trends	If a visibility decrease from 25 to 15 km because of increase in sulfate aerosols in northern mid-latitudes has occurred, then DNA effective UV exposure has decreased in the range 5 to 18% over the last century depending on the height of the planetary boundary layer	Liu et al[76]
Sub-Arctic region	Atmospheric radiation model	Erythemal action spectrum applied to 290-315 nm	Percent change in summer erythemal UV-B exposure rate with change in vertical ozone distribution, aerosol and clouds	Not given; zenith angles 55° and 75°	No UV-B values reported; zenith angle 55°: UV-B increases as much as 0.48 with -20% change in ozone depletion under clear sky; decreases as much as -72% under stratus clouds with no ozone depletion, but 50° increase in troposphere; zenith angle 75%: UV-B increases as much as 68% under stratospheric	Tsay and Stamnes[77]

Site	Model	Measurement	Sampling	Results	Reference
Asian rice climate area; 1200 stations including Songkla, Thailand (7.10°N), Phnom Penh, Cambodia (11.55°N), Calcutta, India (22.4°N), Djakarta, Indonesia (6.11°S), Quezon City, Philippines (14.4°N), Seoul, Korea (37.3°N), Karachi, Pakistan (24.5°N),	Green et al[78] model, and Johnson et al[79] cloud cover model	Daily peak biologically effective UV-B irradiance	Every 30 min, summed over a day, in the middle of each month	H_2SO_4 aerosol, -20% O_3 depletion and 50% increase in troposphere; decreases as much as -70% under stratus clouds with no O_3 depletion; the clear sky radiation amplification factor was found to vary from 2.4 to 3.2 for UV-B, the higher value level being for the plant action spectrum	Batchelet et al[80]
	General plant action spectrum over 280-320 nm			Daily peak values with/without clouds (J m^{-2} day^{-1}), and month, in site sequence shown at left: 6170/10020 (March), 6451/9420 (March/April), 6414/9311 (May/July), 5974/11573 (March/February), 6441/9365 (April/August), 4243/6697 (August/July), 7583/8921 (May/July), 6478/9005 (July), 3410/5661 (August); under 2 x CO_2-induced cloud cover changes, largest change in peak daily UV-B was +17% for Chiang Mai, Thailand (GCM-dependent); highest effective UV-B daily total exposures are up to 13.5 kJ m^{-2} day^{-1} in New Guinea mountains, Himalayas, Tibetan Plateau, Afghanistan and southern tip of India	

Table 1.4. continued

Location(s)	Calculation technique	Actual or weighted effective exposure	UV statistic-definition	Temporal dimensions	Numerical results	Reference
Quiemo, China (38.1°N), Sapporo, Japan (43°N)						
Syowa, Antarctica (69.0°S); Kodaikanal, India (10.14°N); Poona, India (18.32°N); Varanasi, India (25.18°N); New Delhi, India (28.35°N); Srinagar, India (34.05°N)	Green et al[78] model	Actual, 280-340 nm, UV-B and UV-A	Irradiance by wavelength (W m^{-2} nm^{-1}); graphs only	Local noon values; 1987 (O$_3$ data input); monthly for Delhi and Syowa; summer and winter months for remaining locations	No data values given; from visual comparison of graphs it appears that the descending order of UV-B levels are: Kodaikanal (April, 267 DU) = Poona (May, 288 DU) > Delhi (June, 302 DU) = Srinagar (May, 330 DU) > Varanasi (May, 293 DU) > Syowa (November, 255 DU); questionable claims made that November Syowa UV-B levels are "comparable" to summer values at Kodaikanal, but results are insufficient to justify this	Sharma and Srivastava[81]

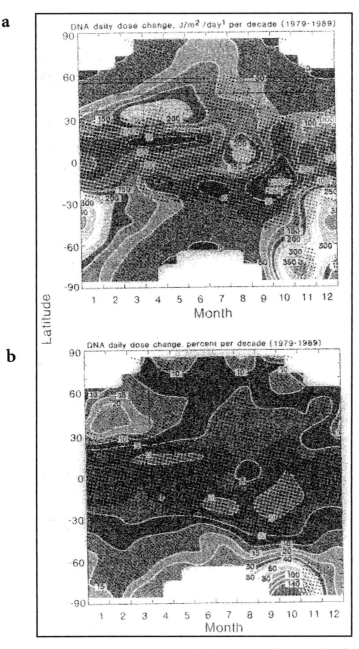

Fig. 1.13a,b. Change in DNA daily dose as J m^{-2} day^{-1} per decade (1979-89) and as percent per decade (1979-89).[31] Reprinted with permission from: Scientific Committee on Problems of the Environment (SCOPE). Effects of Increased Ultraviolet Radiation on Biological Systems. Paris: SCOPE, 1992. © 1992 Scientific Committee on Problems of the Environment (SCOPE). The J m^{-2} day^{-1} values in the upper figure can be converted to mW h m^{-2} by dividing the actual values in the figure by a factor of 3.6.

Table 1.5. Trends (increases) in annual UV values, percent per decade, 1979-89[a]

Latitude	Total O$_3$	RB meter[b]	Erythema[c] induction	Plant damage[c]	DNA damage[c]
85°N	-4.5 ± 3.0	5.1 ± 1.5	4.7 ± 1.4	14.8 ± 4.3	10.1 ± 3.2
75°N	-4.0 ± 2.7	3.9 ± 1.3	4.1 ± 1.4	10.8 ± 3.4	9.0 ± 2.9
65°N	-4.2 ± 1.7	3.4 ± 1.3	4.0 ± 1.5	8.9 ± 3.3	8.1 ± 3.0
55°N	-4.4 ± 1.2	3.3 ± 1.2	4.0 ± 1.5	8.1 ± 3.0	7.7 ± 2.9
45°N	-4.5 ± 1.0	3.1 ± 1.0	4.0 ± 1.4	7.2 ± 2.6	7.2 ± 2.7
35°N	-4.5 ± 1.3	3.0 ± 1.1	4.3 ± 1.6	7.0 ± 2.7	7.5 ± 3.0
25°N	-2.7 ± 1.6	1.8 ± 1.1	2.8 ± 1.9	4.2 ± 2.8	4.8 ± 3.3
15°N	-0.7 ± 1.2	0.5 ± 0.8	0.8 ± 1.4	1.1 ± 1.9	1.3 ± 2.3
5°N	-1.2 ± 1.4	0.8 ± 1.0	1.3 ± 1.6	1.8 ± 2.3	2.2 ± 2.7
5°S	-1.2 ± 1.3	0.8 ± 0.8	1.3 ± 1.4	1.8 ± 2.0	2.1 ± 2.4
15°S	-0.3 ± 1.2	0.2 ± 0.8	0.4 ± 1.3	0.6 ± 1.8	0.7 ± 2.2
25°S	-1.2 ± 1.7	1.0 ± 1.1	1.8 ± 1.7	2.7 ± 2.5	3.3 ± 2.8
35°S	-3.1 ± 1.7	2.3 ± 1.1	3.5 ± 1.6	5.6 ± 2.6	6.3 ± 2.9
45°S	-4.9 ± 1.5	3.7 ± 1.1	5.3 ± 1.6	9.3 ± 2.8	9.9 ± 3.0
55°S	-7.3 ± 1.7	6.0 ± 1.5	7.7 ± 1.9	15.4 ± 4.0	15.1 ± 3.9
65°S	-10.8 ± 2.2	9.5 ± 2.3	11.4 ± 2.8	25.4 ± 6.6	23.4 ± 6.1
75°S	-13.2 ± 2.8	12.8 ± 3.9	15.0 ± 4.8	39.0 ± 13.4	34.4 ± 11.9
85°S	-14.5 ± 3.4	15.6 ± 5.1	16.8 ± 5.7	53.9 ± 20.2	42.7 ± 15.8

[a]Reprinted with permission from: Madronich S. In: Tevini M, ed. UV-B Radiation and Ozone Depletion: Effects on Humans, Animals, Plants, Microorganisms and Materials. Boca Raton, FL: Lewis Publishers, 1993:17-69. © 1993 American Geophysical Union. [b]RB meter: Robertson-Berger integrated erythemal UV measurement device. [c]Biological action spectra used in the correction of the R-B meter data. For descriptions of the spectra, see chapter 2.

UV-B Measurements

Ultraviolet levels have changed with sunspot activity over an 11-year period. We are about halfway past the minimum of this cycle and therefore, the sun can be expected to be emitting more UV again.[10] Concurrent with this periodicity is the normal seasonality of the O$_3$ cycle, with decreases in the winter period of each hemisphere. One study found that, except in the polar regions, an unquestionable decrease in total column O$_3$ by as much as 6.3%, between the years 1983 to 1991 at latitudes above 40°N and S, has not been observed to correspond to increased UV-B levels at the ground.[10] American stations using R-B (Robertson-Berger) erythemal meters have not shown increased UV-B between 1974 and the autumn of 1991. Reasons for this might include the combination of timing in the sunspot cycle, inadequate UV-B monitoring equipment and/or monitoring locations, prevalence of atmospheric sulfate aerosols from volcanic eruptions and from human activity, changes in the frequency of cloud cover and the increase in tropospheric O_3 concentrations.

Measured levels of UV-B and UV-A irradiances for two locations in Europe, north and south, (England and Greece) are also shown in Table 1.3. It is only relatively recently that scientists began UV-B observations in England,[38] or for that matter anywhere in Europe, therefore a long-term measurement database is missing. It will be many years before definitive data are available to indicate what changes are actually occurring. Furthermore, publications reported UV-B irradiance values that are integrated across wavelengths, which is not a useful form of data presentation for plant effects analysis.[40] Such a procedure precludes the necessary weighting of the UV-B irradiances by action spectra across individual wavelengths. This must be done on a spectral basis for broadband measurements, for the hourly or daily results. For additional comments see chapter 2.

At Jungfraujoch, Switzerland (47°N, 7.98°E, 3576 m MSL), using daily totals of flux ratios of Robertson-Berger meter values of UV-B, to total solar radiation (sunburn units, MJ m^{-2}), one study[53] concluded that erythemally weighted UV-B increased by 0.7% ±0.2% per year, if the 1981-89 maximum ratio (0.8) was used. If the 1981-89 mean ratio was used, then the erythemal weighted UV-B increase would be 1.1% ± 0.4% per year. These conclusions would only be valid if the total solar radiation, used as the denominator of the ratio, would not have decreased over that time period. But, the fundamental assumption on which the conclusions were based seems to be violated by the results of another more recent work[11] that shows global solar radiation, as having decreased over several years at several European locations (Table 1.1).

Another well known study published for cities in the U.S., within the latitudes that correspond to some of the southern EEC, revealed declines in erythemal weighted (290-330 nm) UV-B between 1974 and 1985.[36] The largest average trend was -1.1% per year for Minneapolis, MN (44.9°N). However, the validity of the regional representativeness of the 8 locations of the Robertson-Berger meters, with respect to the rest of the nonurban portions of the U.S. has been questioned, because of the increases in urban air pollution over the 12-year period.[82]

The highest value of UV-B observed in Longyearbyen, Svalbard, Norway is about 20% of that at the equator.[48] Specific values for Norway are shown in Table 1.3. In the Southern Hemisphere, at Melbourne, Australia (37.9°S, 145.1°E), during 10-14 December (spring-summer) in 1988, when a 10.5% decrease in total column O_3 occurred, data show a 11.5% increase in UV-B irradiance and 21% increase in effective irradiance.[65] However, the 41-day UV-B irradiance for the total observation period was about the same as during the comparable period in the previous year, because of more cloudiness in 1987-88.

In the Antarctic, the 1988 springtime O_3 hole actually brought summertime DNA-effective UV-B exposures two months early, but the levels did not exceed summertime maximum values.[66] However, the severe 1989 and 1990 O_3 holes, not only brought summertime exposures earlier into the spring season, but the values were also higher than the regular summertime values.

In addition to these latitudinal, there are also altitudinal differences in the fluences of surface level UV-B. Simultaneous measurements of UV irradiance were made from two adjacent sites in Germany (47.5°N, 11.1°E), at Garmisch-Partenkirchen (730 m MSL) and at the Zugspitze (2964 m MSL).[83] Maximum erythemally weighted irradiance levels were observed in Garmisch-Partenkirchen, during March 1996 and June 1995. The March episode was associated with a polar stratospheric cloud, while the previous June episode was associated with an unusually low total O_3 column and broken clouds. The June maximum may be especially significant, because it was preceded by a minimum UV irradiance caused by heavy cloud cover, thus the absolute change in the UV irradiance was very extreme. Even though the maximum irradiance levels increased, these measurements show that the monthly mean spectral UV irradiation in 1995 was not significantly different from the two previous years. This is important, because an increase in the number of extreme fluctuations from low UV irradiation values, followed by very high UV irradiation may be much more significant for the biosphere, than a small, but gradual increase in the average UV dose, because natural adaptation mechanisms may not be able to cope with these extreme fluctuations.

The monthly erythemally weighted irradiation was between 25 and 90% higher at Zugspitze, than at the elevationally lower Garmisch-Partenkirchen site. However, the difference between the two sites cannot be characterized by a single number, because the average monthly ratio of the irradiances was very variable with respect to both the time and the wavelength. The variability in the differences between the two sites indicates that the differences are caused by a combination of several factors, including stray light scattering, cloud effects, air pollutants (e.g., tropospheric ozone), aerosols and albedo. These results show that although UV irradiation measurements from one site (e.g., Garmisch-Partenkirchen) can be used in the assessment of biological effects in its immediate neighborhood, data from a single site are not sufficient to extrapolate even to different altitudes near the measurement site, and certainly should not be used to extrapolate conclusions to the atmospheric chemistry of the global troposphere.

Model UV-B Calculations

Published model calculations of ground level UV and UV-B values are shown in Table 1.4. Even with a decline in total O_3, a theoretical modeling study[32] showed that in the Northern Hemisphere, during the spring and summer, rather than increases, decreases of daily total UV-B and DNA-effective UV-B exposures are possible, because of the increases in tropospheric O_3, due to precursor emissions from industrial sources and biomass burning over the time period 1966-86. The model included partial cloudiness and atmospheric aerosol levels. The same conclusions were found when applying the model to a site in Germany (Hohenpeissenberg, Bavaria) at the summer solstice and on a global basis by latitude throughout the 20 years.

In another study[72] that used a radiative transfer model for latitude bands 30°N-39°N, 40°N-52°N and 53°N-64°N, over a 32 year period (1957-88), no statistically significant trends could be found at any of the latitudinal bands. For shorter time periods, the only significant trend was at 40°N-52°N for 1970-88, where annual integrated mouse carcinoma (MC)-effective UV increased by 3.1% ±1.8% per decade. At specific European sites, only Uccle, Belgium, and Arosa, Switzerland showed statistically significant trends (1970-88) in annual total MC effective UV of +5.06% ±2.86%, and +4.37% ±2.80% per decade, respectively. In comparison, Belsk, Poland and Hohenpeissenberg, Germany did not show such trends.

On an annual basis (without regard for seasonality) over 11 years (1979-89), another modeling study[31] showed that percent change per decade of total plant effective UV-B exposure was 7.0% (±2.7%) to 8.1% (±3.0%) between 35°N-55°N. The curves across global latitudes for this unit of UV-B and for DNA-effective exposure are shown in Figure 1.14. It was concluded that without troposphere O3 pollution and S-aerosols, these values would have been 4% higher. Based on percent per decade changes in annual total plant effective UV-B and DNA-effective UV-B, to changes in annual total column O_3, the latitudinal curves for the respective radiation amplification factors (RAF) are shown in Figure 1.15. On a seasonal basis, according to this study, there was a statistically significant increase in DNA-effective daily exposure within the 30°N-60°N latitude, during late winter and early spring, and north of 50°N, during the summer from 1979 through 1989 (Fig. 1.13a). Also, the largest statistically significant relative increase in DNA-effective daily exposure in the Northern Hemisphere (30°N-50°N) was +20% per decade during December-January (Fig. 1.13b).

Still another model[75] showed a percent change in annual effective UV exposures of +3.1% from 1969-86 for the latitude 40°N-52°N (Table 1.4). An earlier global model of UV-B irradiance[73] showed spring, summer and fall values for Germany of 2.0, 3.2 and 1.0 W m^{-2}, and for the EEC, from south to north by seasons, of 3.0 to 2.0, 3.9 to 2.6, 2.0 to 0.8 W m^{-2}, respectively (Fig. 1.16).

The modeling study of Brühl and Crutzen[32] mentioned previously, also provided global latitudinal results on a seasonal basis. This is of interest with regard to crop ecology. Across the latitudes of Germany (highest crop producer in the world per hectare),[22] for the months of April to September, the 1966 daily noontime UV-B irradiance ranged from 1.6 to 2.2 W m^{-2} (Fig. 1.17). For other latitudes of the EEC, from the north to the south, corresponding values were a little lower or higher. For these same months, the percent change from 1966 to 1986 in daily total UV-B irradiance for Germany and the other EEC countries was shown to be in the range -0.5% to -1.0% (Fig. 1.18). The decline was shown to be attributable to the 60% increase in modeled tropospheric O_3 during the summer in the middle latitudes.[32] However, these results were for daily total UV-B irradiance. When the values for daily total DNA-effective UV-B exposures were analyzed, it was found that from 1966 through 1986, during April through September, the percent change in daily total DNA-effective exposure ranged from about -1.0 in the springtime, to more than

Global Climate and UV-B Radiation 47

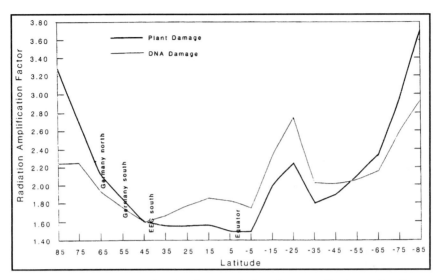

Fig. 1.14. Percent increase per decade of (1979-89) plant damage-effective UV irradiance and DNA damage-effective UV irradiance by latitude.[31]

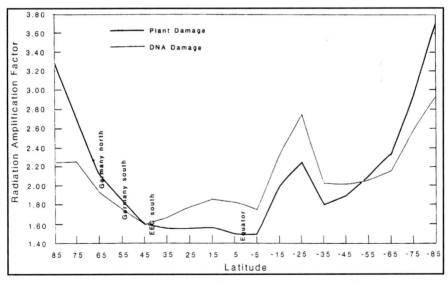

Fig. 1.15. Variations in the Radiation Amplification Factor by latitude, based on percent change per decade of annual total plant-effective UV-B, and DNA-effective UV-B, to total atmospheric O_3.[31]

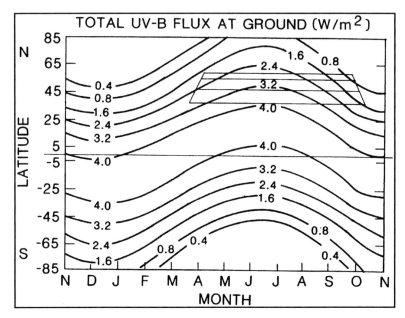

Fig. 1.16. The latitudinal and monthly distribution of UV-B radiation at the ground, computed for clear sky conditions and at a local time of 10:00 a.m. Values include all wavelengths between 280-320 nm.[73]

-2.0, but less than -4.0% (Fig. 1.19) for the latitudes of Germany and the other EEC countries.

One of the modeling studies performed for Athens, Greece,[46] compared measurements of combined UV-B and UV-A against the modeled values with fairly good agreement especially in the summer months (Table 1.4). Because of the scarcity of observed data and, because of the differences between the usefulness of UV-B models for plant ecologists in contrast to meteorologists, Table 1.6 lists some of the models of interest to the former group of scientists.

However, almost all references to UV irradiance values pertain to a horizontal surface. Both human bodies and plants display exposed surfaces at angles other than the horizontal. Such nonhorizontal surfaces will not have the same amount of exposure to UV as the horizontal. A model has been developed based on approximately 800 measurements at 26 different angles at 33 sites for the proportion of incident global UV on inclined surfaces in comparison to UV irradiance on a horizontal plane.[92] The study does not mention the wavelengths but the implication is that they probably ranged from

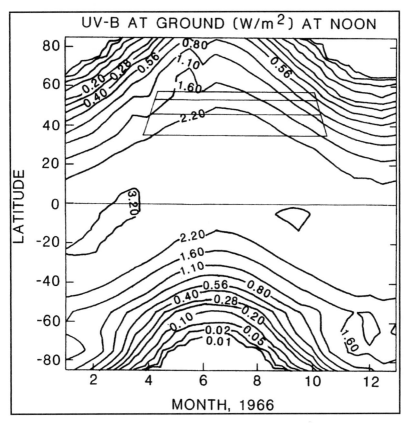

Fig. 1.17. Modeled ground-level UV-B (W m^{-2}) at noon.[32] The boxed area within the upper portion of the figure represents the latitudes of Germany and the European Economic Community.

280 through 400 nm, i.e., both UV-B and UV-A. The tilting effect on irradiance exposure is dependent upon the angle of the surface with respect to the zenith and the UV reflectivity of the surface. Although intended to be applied for the evaluations of human welfare (all components of the environment), in a geometric sense it could also be theoretically applied to crop plants. For example, flat horizontal portions of grass such as the uppermost portions of tillers, will have a relative UV irradiance of 100%, while leaves standing upright will have a relative UV irradiance of only 30%. The problem in an agricultural context is, it appears that there is no compilation of information for the frequency of surface angles of plant parts by crop species or cultivars. Such a study would have to be performed by

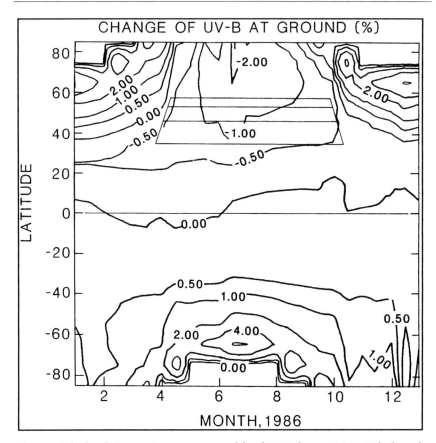

Fig. 1.18. Calculated changes in percent ground-level UV-B from 1966-86.[32] The boxed area within the upper portion of the figure represents the latitudes of Germany and the European Economic Community.

computerized image analysis. In this approach, a limitation would be when the crowding, or plant density reaches a certain level and shading would occur between individuals. Also, other effects such as wind, and changes in the growth form over the life cycle of the crop, would change such angles in an actual field setting.

Spatial Variation Caused by Clouds, Tropospheric Aerosols and O_3

In Table 1.4, the first model listed, and the last four, include the effects of tropospheric O_3, aerosols and cloudiness on the modeled dynamics of UV-B irradiance. Mention has been made of the per-

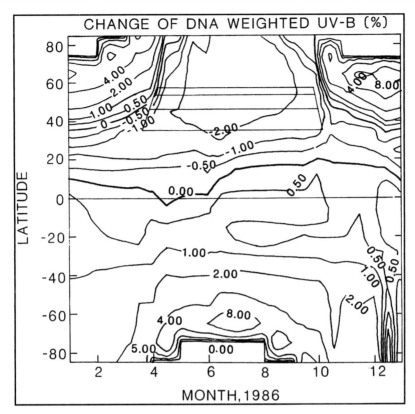

Fig. 1.19. Percent change in DNA-weighted UV-B from 1966-86.[32] The boxed area within the upper portion of the figure represents the latitudes of Germany and the European Economic Community.

cent change in daily total UV-B irradiance over the latitudes of Germany and the other EEC countries from 1966 to 1986, showing a range of -0.5 to -1.0% (Fig. 1.18). That was attributed to the 60% increase in modeled tropospheric O_3 during the summer in the mid-latitudes (Fig. 1.12).[32]

To the north, for the Sub-Arctic region, it was concluded that when the sun was reasonably high in the sky, erythemal UV-B increased by as much as 48% under the clear sky with a -20% depletion of stratospheric O_3.[77] But, it decreased by as much as 72% under stratus clouds with no stratosphere O_3 depletion, with a 50% increase in tropospheric O_3. When the sun was much lower in the sky, UV-B increased by as much as 68% under the conditions of stratospheric

Table 1.6. Mathematical models useful to biologists for simulating ground-level UV irradiance

Required inputs	General structure	Outputs produced	Comparison to measurements	Reference
Elevation, latitude, date, surface albedo, column O_3 thickness, relative aerosol and cloud cover (tenths)	Combination of Green et al[78] model, and Johnson et al[79] cloud cover model	Daily peak plant effective UV-B irradiance	None	Batchelet et al[80]
Latitude, time, date, altitude, surface albedo, total column O_3 concentration and aerosol concentration	Based on param = eterization by Schippnick and Green[84] model of more complex radiative transfer models	Instantaneous spectral direct and diffuse UV-B irradiance (290-320 nm) as mW m^{-2} nm^{-1}	At 7:00 a.m. to 12:00 noon local time, over 3 summer days, at Logan, UT (U.S.) in 1983, visual graphs present calculated spectral irradiance versus measured; no statistical goodness-of-fit results given; author claims model predicts measured values extremely well	Rundel[85]
Latitude, longitude, date, time, total O_3 (optional), environment type, ground cover, air pressure, relative humidity, aerosol level and action spectra choice	Based on Green[44] model	Cloudless instantaneous spectral UV-B irradiance (280-340 nm; W m^{-2} nm^{-1}); no daily totals	Tested for Gainesville, FL (U.S.), Logan, UT (U.S.), Lund, Sweden; USA observed values were within ±20%	Björn and Murphy[86]

Input	Output	Comments	Reference
Zenith angle (date and time), O_3 thickness, aerosol optical depth, ground albedo and ground elevation	Diffuse spectral irradiance expression fit to a series of radiative transfer calculations	At 5 sites over 0°-70°N latitude, 15-4400 m elevation, measured seasonal maximum daily effective UV-B differed by less than 10% (Caldwell et al[87]); at Lauder, New Zealand in summer, spectral irradiance showed ±20%, but DNA irradiance showed ±10% (Seckmeyer and McKenzie[37])	Green,[44] Green et al[78]
Date, time, elevation, snow age parameter, column O_3, precipitable water vapor, turbidity parameters, aerosol absorbance/reflectance ratio, hillslope grade, aspect, view factor, horizon angle, orchard/forest obscurability of sky and direct beam and terrain obscurability of sky	Direct and diffuse spectral irradiance single-layer atmosphere parameterization attenuated by O_3, turbidity and water vapor (not clouds), and corrected for topography and vegetation, with Akima interpolation for wavelength integration and adaptive quadrature method time integration	Spectral and broadband-integrated incoming and net solar radiation at a snow surface in mountains over wavelengths 250 to 5000 nm; W m^{-2}; no demonstrations given for UV-B wavelengths but model should be modifiable for use in snow covered mountain orchards, although no procedure for weighting spectral values by plant action spectra is given	Dozier[88]
		No field tests for specific wavelengths are given; for broadband tests, in the Owens River Drainage of the southern Sierra Nevada, CA, U.S., between 19 March - 16 May 1979, 115 sets of measurements for 280-2800, and 700-2800 nm gave a combined correlation coefficient of 0.994; results significant at 0.99 level	

Table 1.6. continued

Required inputs	General structure	Outputs produced	Comparison to measurements	Reference
Date, time, elevation, snow age parameter, column O_3, precipitable water vapor, turbidity parameters, aerosol absorbance/reflectance ratio, hillslope grade, aspect, view factor, horizon angle, orchard/forest obscurability of sky and direct beam and terrain obscurability of sky	Direct and diffuse spectral irradiance single-layer atmosphere parameterization attenuated by O_3, turbidity and water vapor (not clouds), and corrected for topography and vegetation, with Akima interpolation for wavelength integration and adaptive quadrature method time integration	Spectral and broadband-integrated incoming and net solar radiation at a snow surface in mountains over wavelengths 250 to 5000 nm; W m^{-2}; no demonstrations given for UV-B wavelengths but model should be modifiable for use in snow covered mountain orchards, although no procedure for weighting spectral values by plant action spectra is given	No field tests for specific wavelengths are given; for broadband tests, in the Owens River Drainage of the southern Sierra Nevada, CA, U.S., between 19 March - 16 May 1979, 115 sets of measurements for 280-2800, and 700-2800 nm gave a combined correlation coefficient of 0.994; results significant at 0.99 level	Dozier[88]

Inputs	Model	Results	Reference
Solar angle, direct and diffuse UV-B irradiance above canopy, leaf area index, horizontal leaf distribution (gap factor), leaf structural orientation (leaf angle), leaf optical reflectance transmission spectra, soil UV-B albedo and leaf epidermal transmission spectra	15 layer plant canopy radiation penetration model, based on unpublished modification of Lemon et al[89] SPAM model	UV-B (280-315 nm) spectral irradiances within plant canopy strata (W m^{-2} nm^{-1}); transmission to interior leaf tissue	None	Allen et al[90]
Time, day of year, latitude, longitude, sea level visibility, elevation, albedo and total precipitable water vapor	Chandresekhar radiative transfer for direct and diffuse radiation by wavelength; 11 layer atmosphere extinction procedure; Dave-Furakawa formulation for UV-B and UV-A bands; standard vertical ozone profile with total ozone thickness of 0.34 cm (conceived prior to the late 1970s stratospheric O$_3$ depletion idea)	Clear sky direct and diffuse spectral irradiance at 111 wavelengths from 290-400 nm (not weighted by any action spectra of a receptor)	At unspecified time and date for Davos, Switzerland, at a given UV-B wavelength, calculated spectral direct and diffuse irradiance differed from measured by as much as +40.8% to -23.3%, and +33.1% to -6.9% respectively, but average % difference integrated across 5 wavelengths for direct and diffuse was +6.2% and +4.1% respectively	McCullough and Porter[91]

H_2SO_4 aerosol, 20% depletion of stratosphere O_3 and a 50% increase in the troposphere O_3. However, with no depletion of stratosphere O_3, UV-B decreased by as much as 70% under stratus clouds. Thus, the UV climate is an extremely complex system when all of the intervening factors are considered.

Results from a different modeling study[80] that included cloud effects and ranged over the southeast Asian latitudes of 6°S to 43°N showed that cloud cover had little effect on the seasonality of maximum UV-B exposure at higher latitudes, but did alter the season of maxima in tropical latitudes. Under 2 x concentration of CO_2-induced changes in cloud cover, the largest change in peak daily UV-B was +17% for Chiang Mai, Thailand, but this was dependent on the particular General Circulation Model (GCM) that was used.

At Palmer Station, Antarctica (64.77°S), from 19 September to 21 December 1988 (springtime), it was observed that under average overcast cloudy skies, UV was reduced by about 50%, while under the thickest overcast skies, UV was reduced by about 80%.[93] Seasonally, cloudiness did not affect the increase in UV-B, induced by the depletion of O_3 in the total column, but over hours to days, increased cloudiness could completely compensate for the increase in UV-B. Here, the length of the time period and the timing and duration of ecological processes are important.[83] Short-term biological events that occur over hours to days can be protected from increased UV-B by the duration of cloudiness, if both phenomena occurred over the same time period.

On the basis of an annual average, in the mid-latitudes of the Northern Hemisphere, clouds and air pollution can decrease the erythemal (sunburn of the skin) UV-B irradiance from 22 to 38%.[63] However, under summer conditions, a 10% decrease in monthly mean fractional cloud cover can cause an increase of 1.2 to 6.4% erythemal UV-B irradiance. In contrast, a 10% increase in monthly cloud cover can cause a decrease of 1.2 to 1.6%. The change in erythemal UV-B irradiance associated with total column O_3 change from 1969 to 1986 may be similar to changes in the cloud conditions.[63]

The effect of clouds on UV-B is related to wavelength. Above 300 nm, the ratio of UV to total solar energy flow is not sensitive to cloud scattering, cloud height or surface albedo.[94] It is controlled by cloud thickness, although tropospheric O_3 causes wavelengths smaller than 300 nm to be sensitive to cloud height and surface albedo. However, according to Blumthaler et al[95] the influence of varying cloudiness on the daily totals of global erythemal effective irradiance is considerably greater than the influence of varying ozone. The lowest transmittance of daily totals of the irradiance was 9.4% in Innsbruck, Austria (577 m MSL) and 23.5% at Jungfraujoch, Switzerland (3576 m MSL). According to the authors, the greater transmittance of irradiance at Jungfraujoch was from the smaller thickness of the cloud layer in the high mountains than in the valleys. In this context, cloud type should also be considered. For example, scattering of radiation from the sides of cumulus clouds can enhance total (global) solar irradiance by 20% or more over the maximum solar noon value occurring in the UV-B band.[96,97] In a study in North Carolina, cumulus-type clouds were found to attenuate up to 99% of the incoming UV-B radiation during overcast conditions. However, these same clouds were found to produce localized increases in UV-B (up to 27%) over time scales of < 1 hour under partly cloudy skies when the direct solar beam was unobstructed.

As for tropospheric aerosols, if a decrease in visibility has occurred during the last century from 25 to 15 km, because of an increase in sulfate aerosols in northern mid-latitudes, then DNA-effective UV exposure has decreased in the range of 5 to 18%, depending on the height of the planetary boundary layer.[76] Similar results were derived recently from a study at Mysore, India (12.6°N). The nonmonsoonal springtime daily total UV-B at 283 nm wavelength decreased more so, because of increased atmospheric turbidity resulting from increases in aerosol and dust levels than from changes in total column O_3.[98] However, in this study, a quantitative statistical analysis of the data to support the conclusion was not given. On the other hand, volcanic eruptions can result in increased aerosols in the upper atmosphere and can decrease column O_3 levels, thereby leading to increases in global DNA-effective UV-B exposures.[99] Conversely, it is equally important to note that increases in the levels of tropospheric O_3 and sulfate aerosols can filter the incident UV-B at the surface.[16] Furthermore, according to Kiehl and Briegleb,[100] summer season sulfate aerosol forcing in the Northern Hemisphere completely offsets the greenhouse forcing over the eastern United States and central Europe. This overall discussion shows the difficulty in assessing future changes in the surface level UV-B, particularly in the absence of long-term measured data.

Inaccuracy and Incomplete Processing of UV-B Data

The inaccuracy of the UV-B measurement instruments is one possible source of uncertainty (see chapter 2), but from an engineering perspective there is a constant evolution in improving the scientific techniques. There are other sources of uncertainty, however, that are the result of a failure to plan adequately for a measurement program, or to work in an integrated manner with scientists in related disciplines who might want to make use of the UV-B measurements.

Providing only irradiance values in a graphic form (not a tabular form) by wavelengths which are not weighted by a plant response action spectrum, as is so often done in meteorological studies, makes it difficult to use such measurements for plant effects analysis, by attempting to weight the values after the data have already been published. If given in a tabular form by wavelengths, such data can at least be weighted by a plant action spectrum at a later time to produce plant effective irradiance values. Human welfare oriented UV-B measurements weighted by the erythemal action spectrum, such as from the Robertson-Berger meters, are also relatively useless for analyzing plant responses. If the unweighted irradiance values have been integrated across wavelengths and reported either in a tabular or in a graphic form, it is not possible to use the information for plant effects studies because the integrated values cannot be weighted subsequently by a plant action spectrum.

What is needed from spring through autumn, are at least hourly irradiance values weighted by a plant action spectrum, then integrated across UV-B wavelengths at each hour and subsequently computed for each day to give daily values of plant effective irradiance. These values could then be used to evaluate seasonal changes in plant growth responses on a daily basis. For the analysis of potential crop response to UV-B levels under patterns of cloudiness, tropospheric aerosols and O_3, at current UV-B levels, and for the percent change over the last decade, there is the need for: (1) a 2-dimensional diagram of broadband total daily plant effective UV-B irradiance versus latitude, and (2) a 3-dimensional diagram of broadband hourly plant effective UV-B irradiance versus hour of the day and day of the year, where both diagrams and data sets are truncated in terms of the crop growth season. In middle and northern latitudes, these terms would include truncation during the spring and the autumn by mean dates of last and first frost, and in the frostless subtropics and tropics, truncation by mean last and first dates of seasonal drought. For even greater usefulness, a 3-dimensional map of mean total daily plant effective UV-B irradiance by month, especially April through September in the mid-latitudes, should also be produced on a grid of 0.5 degrees latitude by 0.5 degrees longitude. With such information, crop growth modelers might be able to analyze for the presence or absence of a geographic effect of UV-B irradiance on general crop production patterns, in relation to other climatic and environmental variables. The work of Batchelet et al[80] comes closest to this, but the authors limited their analyses to southeast Asia and did not incorporate their results into a geographic crop growth and production systems model.

Timing and Duration of UV-B Measurements

Frequently, collecting data on a discontinuous daily and seasonal basis during a few weeks from spring through autumn, makes it difficult to evaluate the effects of UV-B on plant growth. Many publications provide data as average values. Few publications provide statistical frequency distributions of the UV-B measurements (ideally, plant effective irradiances). In this context, the only published analysis that we know[101] is flawed by two problems: (1) it describes the data in undefined "relative intensity units", rather than as irradiance values, and (2) it combines both UV-B and low UV-A wavelengths (280-345 nm). In either case, the observations were made at the latitudes of Antarctica, precluding direct relevance for crop growth analysis. If changes do occur in the UV-B climate, certainly biological receptors will be first exposed to changes in infrequent extreme values, not just average values.[102,103]

Another problem arises when measuring ground level UV-B (even if data processing is done as previously suggested), but failing to simultaneously measure ground-level hourly O_3, CO_2 and the more conventional bio-meteorological variables, such as photosynthetically active and global radiation, temperature, precipitation, wind, cloud cover, etc. Without the full set of data on various possible meteorological constraints to plant growth, analysis will continue to be a major problem in sorting the influence of climate on the temporal production patterns in the agricultural landscape.

References

1. German Bundestag. Ozone depletion, changes in UV-B radiation and their effects. In: Protecting the Earth—A Status Report with Recommendations for a New Energy Policy. Bonn, Germany: Deutscher Bundestag, Referat Offentlichkeirsarbeit, 1991:571-616.
2. Scientific Committee on Problems of the Environment (SCOPE). Effects of Increased Ultraviolet Radiation on Biological Systems. Paris: SCOPE, 1992.
3. Leemans R, Cramer WP. The IIASA Database for Mean Monthly Values of Temperature, Precipitation and Cloudiness on a Global Terrestrial Grid. RR-91-18. Laxenburg, Austria: International Institute for Applied Systems Analysis, 1991.
4. Kittel TGF, Rosenbloom NA, Painter TH et al. The VEMAP integrated database for modeling United States ecosystem/vegetation sensitivity to climate change. J Biogeogr 1995; 22(405):857-862.
5. Kittel TGF, Rosenbloom NA, Painter TH et al. The VEMAP Phase I Database: An Integrated Input Dataset for Ecosystem and Vegetation Modeling for the Conterminous United States. http://www.cgd.ucar.edu/vemap. 1996.
6. Olson JS, Watts JA, Allison LJ. Major World Ecosystem Complexes Ranked by Carbon in Live Vegetation: A Database. NDP-017. CDIC Numerical Data Collection. Oak Ridge, TN: Carbon Dioxide Information Center, Oak Ridge National Laboratory, 1985.
7. Sellers PJ, Meeson BW, Hall FG et al. Remote sensing of the land surface for studies of global change: Models—algorithms—experiments. Remote Sens Environ 1995; 51(1):3-26.
8. Jongman RHG. Ecological classification of the climate of the Rhine catchment. Int J Biometeorol 1990; 34:194-203.
9. Crutzen PJ. Ozone depletion—Ultraviolet on the increase. Nature 1992; 356(6365):104-105.
10. Coldiron BM. Thinning of the ozone layer—Facts and consequences. J Amer Acad Dermatol 1992; 27(5):653-662.
11. Stanhill G, Moreshet S. Global radiation climate changes: The world network. Climatic Change 1992; 21:7-75.
12. Németh P, Tóth Z, Nagy Z. Effect of weather conditions on UV-B radiation reaching the earth's surface. J Photochem Photobiol B:Biol 1996; 32:177-181.
13. Boden TA, Kanciruk P, Farrell MP. Trends '90—A Compendium of Data on Global Change. ORNL/CDIAC-36. Oak Ridge, TN: Carbon Dioxide Information Analysis Center, Oak Ridge National Laboratory, 1990.
14. Sladkovic R, Scheel HE, Seiler W. Atmospheric CO_2 records from sites operated by the Fraunhofer Institute for Atmospheric Environmental Research. In: Boden TA, Kaiser DP, Stoss FW, eds. Trends 93: A Compendium of Data on Global Change. ORNL/CDIAC-65. Oak Ridge, TN: Carbon Dioxide Information Analysis Center, Oak Ridge National Laboratory, 1994:148-156.

15. Conway TJ, Tans PP, Waterman LS. Atmospheric CO_2 records from sites in the NOAA/CMDL air sampling network. In: Boden TA, Kaiser DP, Stoss FW, eds. Trends 93: A Compendium of Data on Global Change. ORNL/CDIAC-65. Oak Ridge, TN: Carbon Dioxide Information Analysis Center, Oak Ridge National Laboratory, 1994: 41-119.
16. Frederick JE, Snell HE, Haywood EK. Solar ultraviolet radiation at the earth's surface. Photochem Photobiol 1989; 50:443-450.
17. Gribbin J. Antarctic ozone hole sent ultraviolet levels sky-high. New Sci 1992; 135(1830):17.
18. Frederick JE. Atmospheric ozone and the ultraviolet-radiation environment of the earth. In: Trends in Theoretical Physics, Vol. 2. Redwood City, CA: Addison-Wesley: 1991:123-139.
19. Unsworth MH, Hogsett WE. Combined effects of changing CO_2, temperature, UV-B radiation and O_3 on crop growth. In: Bazzaz F, Sombroek W, eds. Global Climate Change and Agricultural Production. Chichester, England: John Wiley & Sons, 1996:171-197.
20. Stolarski R, Bojkov R, Bishop L et al. Measured trends in stratospheric ozone. Science 1992; 256:342-349.
21. Kerr RA. Ozone destruction worsens. Science 1991; 252:204.
22. Krupa SV, Kickert RN. The Effects of Elevated Ultraviolet (UV)-B Radiation on Agricultural Production. A peer reviewed critical assessment report submitted to the Formal Commission on "Protecting the Earth's Atmosphere" of the German Parliament, Bonn, Federal Republic of Germany, 1993.
23. Herman JR, Hudson R, McPeters R et al. A new self-calibration method applied to TOMS and SBUV backscattered ultraviolet data to determine long-term global ozone change. J Geophys Res 1991; 96(D4):7531-7545.
24. Gleason JF, Bhartia PK, Herman JR et al. Record low global ozone in 1992. Science 1993; 260:523-526.
25. Kerr RA. New assaults seen on earth's ozone shield. Science 1992; 255:797-798.
26. Towe KM. Stratospheric ozone trends—Letters. Science 1992; 257:727-728.
27. Pearce F. Europe exposed to ultraviolet risk as ozone levels hit all time low. New Sci 1992; 134(1816):5.
28. Frederick JE. Trends in atmospheric ozone and ultraviolet-radiation—Mechanisms and observations for the Northern Hemisphere. Photochem Photobiol 1990; 51(6):757-763.
29. Austin J, Butchart N, Shine KP. Possibility of an Arctic ozone hole in a doubled-CO_2 climate. Nature 1992; 360:221-225.
30. Mahlman JD. A looming Arctic ozone hole? Nature 1992; 360:209-210.
31. Madronich S. Implications of recent total atmospheric ozone measurements for biologically-active ultraviolet-radiation reaching the earth's surface. Geophys Res Lett 1992; 19(1):37-40.

32. Brühl C, Crutzen PJ. On the disproportionate role of tropospheric ozone as a filter against solar UV-B radiation. Geophys Res Lett 1989; 16(7):703-706.
33. Hood LL, McCormack JP. Components of interannual ozone change based on Nimbus-7 TOMS data. Geophys Res Lett 1992; 19(23):2309-2312.
34. Thompson AM. The oxidizing capacity of the earth's atmosphere: Probable past and future changes. Science 1992; 256:1157-1165.
35. Garadzha MP, Nezval YI. Ultraviolet radiation in large cities and possible ecological consequences of its changing flux due to anthropogenic impact. In: Proceedings of the Symposium on Climate and Human Health. WPCA Report 2. Leningrad: World Climate Programme Applications, 1987:64-68.
36. Scotto J, Cotton G, Urbach F et al. Biologically effective ultraviolet radiation: Surface measurements in the United States, 1974 to 1985. Science 1988; 239:762-764.
37. Seckmeyer G, McKenzie RL. Increased ultraviolet-radiation in New Zealand (45° S) relative to Germany (48° N). Nature 1992; 359(6391):135-137.
38. Webb AR. Spectral measurements of solar ultraviolet-B radiation in southeast England. J Appl Meteorol 1992; 31(2):212-216.
39. Webb AR. Ultraviolet-radiation—The missing link in the ozone debate. Norsk Geolog Tidsskr 1991; 71(3):211-213.
40. Webb AR. Solar ultraviolet-B radiation measurement at the earth's surface: Techniques and trends. In: Abrol YP, Govindjee, Wattal PN et al, eds. Impact of Global Climatic Changes on Photosynthesis and Plant Productivity. New Delhi, India: Oxford & Ibh Publ. Co. PVT. Ltd., 1991:23-37.
41. Webb AR. Solar ultraviolet radiation in southeast England: The case for spectral measurements. Photochem Photobiol 1991; 54(5):789-794.
42. Jokela K, Leszczynski K, Visuri R et al. Increased UV exposure in Finland in 1993. Photochem Photobiol 1995; 62:101-107.
43. Bird R, Riordan C. Simple solar spectral model for direct and diffuse irradiance on horizontal and tilted planes at the earth's surface for cloudless atmospheres. J Clim Appl Meteorol 1986; 25:87-97.
44. Green AES. The penetration of ultraviolet radiation to the ground. Physiol Plant 1983; 58:351-359.
45. Leckner B. The spectral distribution of solar radiation at the earth's surface elements of a model. Solar Energy 1978; 20:143-150.
46. Katsambas A, Andoniou C, Stratigos J et al. A simple algorithm for simulating the solar ultraviolet-radiation at the earth's surface—An application in determining the minimum erythema dose. Earth, Moon & Planets 1991; 53(3):191-204.
47. Zerefos CS, Meleti C, Bais AF et al. The recent UVB variability over southeastern Europe. J Photochem Photobiol B: Biol 1995; 31:15-19.
48. Henriksen K, Claes S, Svenoe T et al. Global, spectral ultraviolet and visible measurements in the Arctic. Norsk Geolog Tidsskr 1991; 71(3):191-193.

49. Henriksen K, Stamnes K, Østensen P. Measurements of solar U.V., visible and near I.R. irradiance at 78°N. Atmos Environ 1989; 23(7):1573-1579.
50. Stamnes K, Henriksen K, Østensen P. Simultaneous measurement of UV radiation received by the biosphere and total ozone amount. Geophys Res Lett 1988; 15(8):784-787.
51. Blumthaler M, Ambach W. Spectral measurements of global and diffuse solar ultraviolet-B radiant exposure and ozone variations. Photochem Photobiol 1991; 54(3):429-432.
52. Ambach W, Blumthaler M, Wendler G. A comparison of ultraviolet-radiation measured at an arctic and an alpine site. Solar Energy 1991; 47(2):121-126.
53. Blumthaler M, Ambach W. Indication of increasing solar ultraviolet-B radiation flux in alpine regions. Science 1990; 248:206-208.
54. Dehne K. Design and performance of a new instrument for measuring UV-B global radiation. In: WMO Technical Conf. Instruments and Methods of Observation (TECIMO), 27-30 July 1977, Hamburg. WMO No. 480. Geneva, Switzerland: WMO, 1977:173-178.
55. Behr HD. Net total and UV-B radiation at the sea surface. J Atmos Chem 1992; 15:299-314.
56. Kushelevsky AP. Further solar UV spectral measurements at the Dead Sea. In: Riklis E, ed. Photobiology. New York: Plenum Press, 1991:1023-1026.
57. Akrawi A, Ahmed B. Fluctuations of global ultraviolet-radiation under cloudless conditions in Iraq. Clean and Safe Energy Forever 1990:2176-2180.
58. Kollias N, Baqer AH, Sadiq I. Measurements of solar middle ultraviolet radiation in a desert environment. Photochem Photobiol 1988; 47(4):565-569.
59. Kerr JB, McElroy CT. Evidence for large upward trends of ultraviolet-B radiation linked to ozone depletion. Science 1993; 262:1032-1033.
60. Correll DL, Clark CO, Goldberg B et al. Spectral ultraviolet-B radiation fluxes at the earth's surface—Long-term variations at 39° N, 77° W. J Geophys Res-Atmos 1992; 97(D7):7579-7591.
61. Mims III FM, Ladd JW, Blaha RA. Increased solar ultraviolet-B associated with record low ozone over Texas. Geophys Res Lett 1995; 22:227-230.
62. Goldberg B. The solar ultraviolet—A brief review. In: Boer KW, ed. Advances in Solar Energy. New York, NY: Plenum Press, 1986: 357-386.
63. Frederick JE, Snell HE. Tropospheric influence on solar ultraviolet radiation: The role of clouds. J Clim 1990; 3:373-381.
64. Lee DW, Downum KR. The spectral distribution of biologically-active solar radiation at Miami, Florida, USA. Int J Biometeorol 1991; 35(1):48-54.
65. Roy CR, Gies HP, Elliott G. Ozone depletion. Nature 1990; 347: 235-236.

66. Lubin D, Mitchell BG, Frederick JE et al. A contribution toward understanding the biospherical significance of antarctic ozone depletion. J Geophys Res - Atmos 1992; 97(D8):7817-7828.
67. Ryan KG, Smith GJ, Rhoades DA et al. Erythemal ultraviolet insolation in New Zealand at solar zenith angles of 30° and 45°. Photochem Photobiol 1996; 63:628-632.
68. Joseph JH, Wiscombe WJ, Weinman JA. The delta-Eddington approximation for radiative flux transfer. J Atmos Sci 1976; 33:2452-2459.
69. Toon OB, McKay CP, Ackerman TP et al. Rapid calculation of radiative heating rates and photodissociation rates in inhomogeneous multiple scattering atmospheres. J Geophys Res 1989; 94:16287-16301.
70. WMO (World Meteorological Organization). Scientific Assessment of Stratospheric Ozone, Vol. I, Global Ozone Research and Monitoring Project. Report No. 20. Geneva, Switzerland: World Meteorological Organization, 1989.
71. Frederick JE, Lubin D. The budget of biologically active ultraviolet radiation in the earth atmosphere system. J Geophys Res 1988; 93:3825-3832.
72. Frederick JE, Weatherhead EC, Haywood EK. Long-term variations in ultraviolet sunlight reaching the biosphere—Calculations for the past 3 decades. Photochem Photobiol 1991; 54(5):781-788.
73. Serafino GN, Frederick JE. Global modeling of the ultraviolet solar flux incident on the biosphere. In: Hoffman JS, ed. Assessing the Risks of Trace Gases That Can Modify the Stratosphere, Vol. VII, Technical Support Documentation—Atmospheric Science Papers. EPA 400/1-87/001G. Washington, DC: U.S. Environmental Protection Agency, 1987.
74. Stamnes K, Tsay S, Wiscombe W et al. Numerical stable algorithm for discrete-ordinate-method radiative transfer in multiple scattering and emitting layered media. Appl Optics 1988; 27(12):2502-2508.
75. Dahlback A, Henriksen T, Larsen SHH et al. Biological UV doses and the effect of an ozone layer depletion. Photochem Photobiol 1989; 49(5):621-625.
76. Liu SC, Mckeen SA, Madronich S. Effect of anthropogenic aerosols on biologically-active ultraviolet-radiation. Geophys Res Lett 1991; 18(12):2265-2268.
77. Tsay SC, Stamnes K. Ultraviolet-radiation in the Arctic—The impact of potential ozone depletions and cloud effects. J Geophys Res - Atmos 1992; 97(D8):7829-7840.
78. Green AES, Cross KR, Smith LA. Improved analytic characterization of ultraviolet skylight. Photochem Photobiol 1980; 31:59-65.
79. Johnson FS, Mo T, Green AES. Average latitudinal variation in ultraviolet radiation at the earth's surface. Photochem Photobiol 1976; 23:179-188.
80. Batchelet D, Barnes PW, Brown D et al. Latitudinal and seasonal variation in calculated ultraviolet-B irradiance for rice growing regions of Asia. Photochem Photobiol 1991; 54(3):411-422.

81. Sharma MC, Srivastava BN. Ultraviolet-radiation received in Antarctica in comparison with the Indian Region. Atmos Environ 1992; 26(4):731-734.
82. Grant WB. Global stratospheric ozone and UV-B radiation. Science 1988; 242:1111-1112.
83. Seckmeyer G, Mayer B, Bernhard G et al. New maximum UV irradiance levels observed in Central Europe. Atmos Environ 1997; 31:2971-2976.
84. Schippnick PF, Green AES. Analytical characterization of spectral actinic flux and spectral irradiance in the middle ultraviolet. Photochem Photobiol 1982; 35:89-101.
85. Rundel R. Computation of spectral distribution and intensity of solar UV-B radiation. In: Worrest RC, Caldwell MM, eds. Stratospheric Ozone Reduction, Solar Ultraviolet Radiation and Plant Life. Berlin: Springer-Verlag, 1986:49-62.
86. Björn LO, Murphy TM. Computer calculation of solar ultraviolet radiation at ground level. Physiol Veg 1985; 23(5):555-561.
87. Caldwell MM, Robberrecht R, Billings WD. A steep latitudinal gradient of solar ultraviolet-B radiation in the arctic alpine life zone. Ecology 1980; 61:600-611.
88. Dozier J. A clear sky spectral solar radiation model for snow-covered mountainous terrain. Water Resources Res 1980; 16(4):709-718.
89. Lemon E, Stewart DW, Shawcroft RW. The sun's work in a corn field. Science 1971; 174:371-378.
90. Allen Jr LH, Gausman HW, Allen WA. Solar ultraviolet radiation in terrestrial plant communities. J Environ Qual 1975; 4(3):285-294.
91. McCullough EC, Porter WP. Computing clear day solar radiation spectra for the terrestrial ecological environment. Ecology 1971; 52(6):1008-1015.
92. Schauberger G. Model for the global irradiance of the solar biologically-effective ultraviolet-radiation on inclined surfaces. Photochem Photobiol 1990; 52(5):1029-1032.
93. Lubin D, Frederick JE. The ultraviolet radiation environment of the Antarctic Peninsula: The roles of ozone and cloud cover. J Appl Meteorol 1991; 30:478-493.
94. Spinhirne JD, Green AES. Calculation of the relative influence of cloud layers on received ultraviolet and integrated solar radiation. Atmos Environ 1978; 12:2449-2454.
95. Blumthaler M, Ambach W, Cede A et al. Attenuation of erythemal effective irradiance by cloudiness at low and high altitude in the alpine region. Photochem Photobiol 1996; 63:193-196.
96. Mims III FM, Frederick JE. Cumulus clouds and UV-B. Science 1994; 371:291.
97. Schafer JS, Saxena VK, Wenny BN et al. Observed influence of clouds on ultraviolet-B radiation. Geophys Res Lett 1996; 23:2625-2628.
98. Prasad BSN, Gayathri HB, Krishnan NM et al. Seasonal-variation of global UV-B flux and its dependence on atmospheric ozone and

particulate at a low latitude station. Tellus Ser B—Chem Phys Meteorol 1992; 44(3):237-242.
99. Vogelmann AM, Ackerman TP, Turco RP. Enhancements in biologically effective ultraviolet-radiation following volcanic eruptions. Nature 1992; 359(6390):47-49.
100. Kiehl JT, Briegleb BP. The relative roles of sulfate aerosols and greenhouse gases in climate forcing. Science 1993; 260:311-314.
101. Barysheva VI, Lebedinets VN, Ogurtsov VI. Analysis of the earth's ultraviolet-radiation variability from satellite measurements and model calculations. Adv Space Res 1993; 13(1):385-388.
102. Ausubel JH. A second look at the impacts of climate change. Amer Sci 1991; 79:210-221.
103. Katz RW, Brown BG. Extreme events in a changing climate: Variability is more important than averages. Climatic Change 1992; 21:289-302.

CHAPTER 2

Assessment Methodology

Introduction

Any assessment of the effects of elevated UV-B radiation on crops has four experimental components: (a) quantification of the UV-B level itself in various treatments, including the ambient; (b) use of an exposure system which will clearly meet the stated objective; (c) definition of the exposure dose-response dynamics; and (d) reproducible measurements of the crop response that can result in the establishment of cause-effect relationships. This chapter deals with the first three components and the observed effects on crops themselves are discussed in subsequent chapters.

Instrumentation for UV-B

UV-B monitoring instruments can be classified as: (a) broad-band, cumulative and (b) narrow-band, high resolution or spectral. Webb[1] has provided an excellent summary of the techniques for the measurement of UV-B radiation. Among these techniques, the most widely used instruments are of the broad-band type, which provide values of total irradiation across the UV-B band, often weighted by the sensor response function to give an "equivalent" or "effective" measure of the radiation. Such a measure is frequently approximated to a biological response function as in the case of Robertson-Berger Erythemal Meter (Table 2.1). However, in the case of UV-B, tailoring the response of the instrument by the choice of the detector and/or filters is both difficult and critical. Figure 2.1 shows the generalized action spectra for: human erythema, plant damage, DNA damage, the response of the Robertson-Berger meter and the shape of the solar (parts of UV-B) spectrum.[1] Because of the steep and opposing slopes of the solar spectrum and biological responses across the UV-B band, any mismatch of the instrument response and the biological

---

Elevated Ultraviolet (UV)-B Radiation and Agriculture, by Sagar V. Krupa, Ronald N. Kickert and Hans-Jürgen Jäger.
© 1998 Springer-Verlag and Landes Bioscience.

Table 2.1. A comparative summary of the characteristics of different UV-B measurement instruments[a]

Instrument type	Measurement principle	Measurement uncertainty
Simple		
RM2	2-channel photodiode receiver with filters	8% calibration uncertainty
Impact-related[a]		
Robertson–Berger Spectrometer (sunburn UV meter)[b]	Selective transformation of UV radiation into visible radiation by use of phosphorus and detection by photodiodes	Not known
MOH-type[c]	Simulation of the long-wave portion of the erythema (sunburn) threshold effect function by the use of special interference filter combination and detection by UV sensitive, temperature controlled photomultiplier	10% for the summer readings (excluding calibration lamp problems)
Narrow-band		
Brewer Ozone Spectrophotometer	Single-grid monochromator with $NiSO_4$ pre-filter and scanning of the spectral region between 290 and 320 nm through optical grid adjustment	Not known
Optronics Model 742 High Accuracy UV-Visible Spectroradiometer	Dual-grid monochromator (200-800 nm) and detection by (S-20 cathode) UV photomultiplier, computer controlled unit	Not known
LI-1800 Portable Spectroradiometer	Single-grid monochromator and pre-filter selector for spectral region from 300 to 850 nm; bandwidth 4 or 2 nm (upon request); scanning time for the total spectrum ~20 seconds, detection by silicon diode; computer controlled unit	Calibration accuracy at 300 nm is ± 10%

[a]Impact-related, refers primarily to human sunburn or erythema and not to photoreactive plant responses. Further, these instruments do not provide spectral data for individual wavelengths. Different plant processes exhibit different degrees of sensitivity to different wavelengths of UV-B. [b]There are other similar instruments that are commercially available. [c]Meteorological Observatory, Hamburg, Germany.

Assessment Methodology

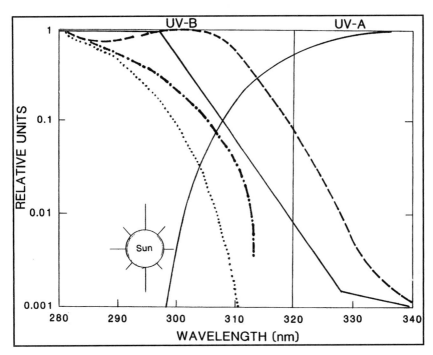

Fig. 2.1. The generalized action spectra normalized at 280 nm for erythema (———), generalized plant damage (— • — • —) and DNA (• • • •), the response spectrum of the Robertson-Berger meter (— — — — —) and the shape of the solar (UV-B) spectrum.[1]

response that it is considered to represent, could greatly reduce the usefulness of the information from such measurements.[1] This problem is compounded when the instrument responds to larger wavelengths past the biological response. Such detection of biologically ineffective (in terms of adverse effects) longer wavelength radiation can mask the changes at shorter wavelengths, which may be very small in energy terms, but very effective in biological terms.

The broad-band radiometers are frequently in use because they are relatively cheap, small and easy to operate. Generally they are also very sturdy and thus can be used in long-term field measurements. In contrast to the broad-band instruments, narrow-band methods have the advantage of providing spectral data. A model output[3] predicted, for example, that with overhead sun and typical O_3 amounts, a 10% decrease in column O_3 would result in a 20% increase in UV-B penetration at 305 nm, a 250% increase at 290 nm and a 500% increase at 287 nm, all within the UV-B band. Such a prediction should by tempered by the fact that the path of UV-B at the surface is diffuse, in comparison to its direct path in the stratosphere.[2] Tropospheric O_3 and aerosols have the same property of filtering UV-B as stratospheric O_3.[4,5] However, O_3 at the surface is phytotoxic[6] and acid aerosols can increase the phytotoxic effects of O_3.[7] Because of these and other important considerations, many nations are attempting to reduce NO_x and SO_x emissions,[8] the precursors to O_3 and fine particle aerosols. Such efforts could result in increased UV-B at the surface, and since different plant processes respond differently to different wavelengths of UV-B, detailed spectral data are vital for reducing the uncertainties of predicting crop responses to present and future ambient UV-B levels.

According to Webb,[1] good solar spectral UV-B measurements require high performance instruments. The shape of the solar spectrum across the UV-B band imposes a need for good wavelength resolution and precision, excellent stray light rejection, high sensitivity and dynamic range of several orders of magnitude. These criteria are satisfied by spectroradiometers employing double gratings to separate radiation into its component wavelengths and then a suitable detector (often a photomultiplier tube) to measure spectral intensities (refer to Table 2.1).

A different approach for detecting radiation spectra is currently employed in instruments used for measuring gases such as O_3 in the atmosphere, by taking the ratio of two wavelengths of radiation, one wavelength absorbed by the gas and the other one not absorbed by the gas. Such instruments use an array of diodes to obtain a "snapshot" of the radiation spectrum, which can subsequently be split into component wavelengths and their intensities recorded (e.g., modified Brewer Spectrophotometer). On the negative side, the measurement uncertainty (e.g., artifacts caused by the detection of stray light) associated with such instruments is not fully known at the present time (Table 2.1). On the positive side, such instruments can provide data on both the atmospheric gas (e.g., O_3) concentration and radiation. There is a critical need for such information, since, for example, both O_3 and UV-B are of concern in the context of future crop production. At the present time, elevated or episodal levels of surface O_3 and UV-B are considered to occur sequentially[9] and therefore, their individual frequency of occurrence during the crop growth season and their impacts constitute another major source of uncertainty in conducting relevant, realistic, location-specific experiments.

Gardiner et al[10] completed an intercomparison of the performance of six different UV spectroradiometers (Table 2.2). The following is their summary quoted verbatim:

"Accurate spectral measurements of solar ultraviolet radiation are essential to an understanding of the photochemical and biological effects of ozone depletion. The analysis of spatial and temporal variations in ultraviolet fluxes will depend on the collation of spectra from many independent laboratories. However, results from diverse instruments operated in isolation may not be consistent with each other. To investigate the compatibility of different designs of the spectroradiometer, a blind trial of six distinct instrument types was carried out at a suburban site in Greece. Comparisons were performed in daylight, and with tungsten lamps indoors. Excellent agreement was obtained in the relative spectral response of the instruments, but their absolute lamp calibrations varied, and did not generally agree with results from the daylight experiments. Simultaneous spectral scans by all instruments revealed discrepancies attributable to stray light, bandwidth and cosine response, which would not otherwise have been apparent."

Most recently, Bernhard and Seckmeyer[11] have developed a new instrumental radiation path entrance optics for solar spectral measurements. This entrance optics reduced the uncertainties associated with the cosine response to ±4%. Solar spectroradiometric measurements carried out with the new optics were compared with the simultaneous measurements with a second spectroradiometer (conventional diffuser) and the deviations of up to 12% between the two systems, were quantitatively explained to within 3%.

Methodology for the Exposure of Crops

As with the previous discussion, Webb[1] has provided an excellent analysis of the considerations relevant to the artificial exposures of crops to elevated UV-B levels. The lack of comprehensive knowledge of the effects of UV-B, particularly under field conditions, and the limitations in our understanding of the mechanisms of its action and joint effects with other environmental variables, has been largely due to the difficulty in conducting realistic experiments under ambient conditions.[9] Plants need light to grow, and therefore the present day ambient light regimes are the control at a given location. Plants are also sensitive, depending on the responses of their different growth processes, to different wavelengths of light.[1] Thus, it is

Table 2.2. Instrument specifications in the European intercomparison of ultraviolet spectroradiometers[a,b]

	AI	AW	B	GB	GR	N
Spectrometer type	Bentham DM150	JY DH10	JY DH10	Optronics 742	Brewer MK II	JY HR320
Focal length/nm	150	100	100	100	160	320
Gratings	Two	Two	Two	Two	One	One
- lines/mm	2400	1200	1200	1200	1800	1200
Bandwidth (FWHM)/nm	1.0	1.4	0.4	1.5	0.6	0.04
Sampling interval/nm	0.5	1.0	0.17	1.0	0.5	0.1
Spectral range/nm						
- from	290	280	250	280	290	290
- to	500	400	370	400	325	600
Scan duration/s	180	120	192	235	185	390
Diffuser	Teflon	Integ. sphere	Double quartz	Teflon	Teflon	Quartz
Weatherproof	No	No	No	No	Yes	Yes
Automatic	Yes	No	Yes	No	Yes	Yes

[a]Reprinted with permission from: Gardiner BG, Webb AR, Bais AF et al. Environ Technol 1993; 14:25-43. © 1993 Selper Publications. [b]AI, Austria - Innsbruck - Institut für Medizinische Physik; AW, Austria - Wien Universität für Bodenkultur, Vienna; B, Belgium - Institut d'Aéronomie Spatiale de Belgique; GB, Great Britain University of Reading; GR, Greece - University of Thessaloniki; N, Norway University of Tromsø.

the ratio of different wavelengths that is critical in evaluating the effects of radiation on plant growth, function and productivity. For example, under low light intensity, many plant species respond adversely to elevated UV-B levels.[5,12] A plant species not saturated by photosynthetically active radiation may suffer through a corresponding evolutionary expectation that other radiation levels are also low and defense mechanisms against UV-B (e.g., foliar accumulation of UV-B absorbing pigments) do not need to be fully developed. In contrast, under high levels of visible light, such as under light intensities of the ambient, there can be photorepair to elevated UV-B levels.[12] Furthermore, even in artificial UV-B supplementation experiments, added UV-A radiation and other effects associated with energized lamps can significantly affect plant response in outdoor experiments.[13]

Almost all of the UV-B-plant response studies have used artificial UV radiation sources (Fig. 2.2) to enhance the short wavelength portion of an already existing artificial or natural source of light. However, the spectral distribution of the artificial sources is not the same as that of the sun and therefore, their relative effectiveness has to be assessed by using a biological weighting function. The DNA-action spectrum[14] or the generalized plant damage action spectrum[15] has often been used, but neither of these is fully representative of the response of whole plants of different species.[16] For additional information on this aspect, see the section below.

Because of the spectral properties of UV lamps, filters are commonly used to absorb the UV-C and short wavelength UV-B (the actual reason for this is not completely clear, in light of Cutchis,[3] other than to conclude that it is based on the lamp and filter characteristics) radiation emitted by these lamps (Fig. 2.2). Most investigators in Europe have used lamps manufactured by Phillips of The Netherlands coupled with glass cutoff filters (Schott WG Series).[17] Tevini et al[18] used a O_3 flow cell as a filter for natural solar radiation. This approach is highly desirable, if a functional system can be fully deployed, since it simulates most of the natural filtration system of the ambient O_3.[3] In comparison to all these filtration systems, most scientists in North America have used Westinghouse, U.S., lamps coupled with cellulose acetate plastic film. Krizek et al[19] used UV-B-313 lamps manufactured by Q-Panel Company (Cleveland, U.S.),

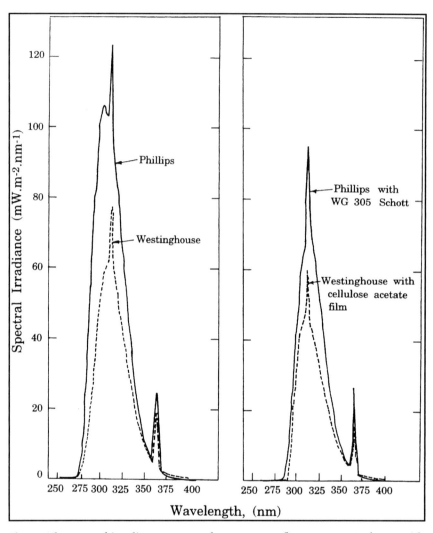

Fig. 2.2. The spectral irradiance at 30 cm from common fluorescent UV-B lamps with and without filters. The lamps and filters were new and the spectral irradiance was measured with a double monochromator spectroradiometer (Optronic Laboratories). The Phillips TL 40W/12 and the Westinghouse FS40 lamps without filters are shown on the left and with filters on the right. Schott WG 305 sharp cutoff absorption filters (2 mm thickness) were used with the Phillips lamp and cellulose acetate plastic film (0.13 mm thickness) with the Westinghouse lamp. These lamp/filter combinations have commonly been used in studies in Europe and North America, respectively. Reprinted with permission from: Caldwell MM, Camp LB, Warner CW et al. In: Worrest RC, Caldwell MM, eds. Stratospheric Ozone Reduction, Solar Ultraviolet Radiation and Plant Life. Berlin: Springer-Verlag, 1986:87-111. © 1986 Springer-Verlag.

coupled with cellulose diacetate or polyester filters. A concern here is the possible effect of the aging of the filter (cellulose acetate) after continued use.[20]

Tables 2.3 and 2.4 provide a comparative summary of the characteristics of various exposure systems. When the lamp systems are used to supplement normal solar radiation, the discrepancy between the spectral irradiance received from the sun and the lamp is much less compared to the growth chamber experiments. However, since the filtered fluorescent lamps emit longer wavelength radiation than is desired for the supplement, it is necessary to use control lamps which are identical except, they are filtered to exclude the UV-B radiation. Thus, the amount of UV-A radiation in the control will be the same as that from the treatment lamps. However, this would also account for microclimatic changes such as the shading effect from the treatment lamp system. Frequently Mylar (DuPont Company, Newark, U.S.) filters are used to provide the control radiation for the UV-B treatment (Fig. 2.3).

In all these artificial lamp exposure systems, the fluence of UV-B at the plant canopy is simply varied by altering the distance between the height of the lamp bank and the plant canopy. In contrast, Caldwell et al[39] developed an exposure system which allows: (a) lamp emittance to be modulated over a large dynamic range (50:1); (b) stable lamp operation starting at low temperatures and a sensitive feedback loop to compensate for both atmospheric conditions such as cloud cover; as well as (c) changes in radiation emittance from the lamps due to factors such as temperature and lamp age (Fig. 2.4). To adjust for some of the artifacts introduced by the lamp age and change in emittance, most investigators pre-solarize UV lamps for about 100 hours. Table 2.4 provides a summary of the modulated exposure systems.

UV-B Dosimetry

In the absence of reproducible measured and demonstrable total column ozone reduction at all latitudes, the solar spectrum increases rapidly with increasing wavelength in the UV region. Conversely, absorption by biomolecules increases with decreasing wavelength.[5] This is the scenario predicted by Cutchis[3] for a loss in total column ozone and change in UV-B, by decreasing wavelength. Therefore, the interest is not so much in the total UV-B irradiance, but in the effectiveness of that irradiance in initiating some photobiological effect.

Table 2.3. Switched or square-wave outdoor supplementation systems for studies of increased UV-B radiation[a,b]

Reference	Lamp type	High/low frequency	Filters Treatment	Control	Species studied	Description
Johanson et al;[21,22] Gehrke et al[23]	Q-Panel UVB-313	LF	CA plus Plexiglas (Röhm 2458)	Window glass	Subarctic heathland; dwarf shrubs including decomposition	Timer controlled in two steps to simulate 15% O_3 depletion at Abisko, Sweden, calculated using the model of Björn and Murphy.[24] Duration adjusted biweekly. Also with CO_2 in small open-top chambers.
Gehrke et al[25]	Philips TL/4W	LF	CA	PE 'Mylar-S'	Peatland ecosystem dominated by *Sphagnum fuscum*	Timer controlled to simulate 15% O_3 depletion at Abisko, Sweden, calculated using the model of Björn and Murphy.[24] Duration of adjusted biweekly.
Teramura et al[26]	Westinghouse FS-40	LF	CA	PE 'Mylar'	*Glycine max*	Timer controlled in 2 steps for 6 h daily to simulate 16 and 25% O_3 depletion at Beltsville, U.S., calculated using the model of Green et al.[27] Different dose by varying distance from lamps.

Reference	Lamp			Species	Methodology
Sullivan and Teramura;[28] Sullivan et al[29]	Q-Panel UVB-313	LF	CA	PE	*Pinus taeda; Liquidambar styraciflua*	Timer controlled for 6 h per day to simulate 16 and 25% O$_3$ depletion at Beltsville, U.S., calculated using the model of Green et al.[27] Different dose by varying distance from lamps.
Booker et al;[30] Fiscus et al[31]	Q-Panel UVB-313	LF	CA	PE	System description; *Glycine max*	Inside PVC open-top chambers, timer controlled, flux and duration adjusted biweekly, manual intervention if overcast; 3 year study with gaseous O$_3$ interaction simulating a 15, 20 and 35% O$_3$ depletion at Beltsville, U.S., calculated using the model of Green et al.[27]
Petropoulou et al[32]	Philips TL40/12	LF	CA	LNE	Mediterranean pines	Timer controlled exposure in 2 steps to simulate 15% O$_3$ depletion at Patras, Greece, calculated using model of Björn and Murphy.[24] Daily exposure duration adjusted monthly.

Table 2.3. continued

Reference	Lamp type	High/low frequency	Filters Treatment	Control	Species studied	Description
H. Ro-Poulsen (pers. comm.); Zeuthen[33]	Philips TL40/12	LF	CA	Window glass	*Fagus sylvatica*; *Quercus robur*	Field study near Copenhagen, Denmark on 6 year old, 1 m high trees. Simulated 15 and 30% O_3 depletion using model of Björn and Murphy.[24]
H. Ro-Poulsen (pers. comm.)	Philips TL20/12	LF	CA	Window glass	*Cladonia mitis*	Field study at Qaanaaq, Greenland. Simulated 15 and 30% O_3 depletion using model of Björn and Murphy.[24]
Dai et al;[34] Olszyk et al[35]	Q-Panel UVB-313	LF	CA	PE 'Mylar'	Rice (*Oryza sativa*) cultivars	Field study at IRRI, Philippines using 6 h switched treatment to simulate 20% O_3 depletion.
Rozema et al[36]	Philips TL40/12	HF	CA	PE 'Mylar'	Crop species, species of natural	Switched exposures. Modulated system

Zipoli and Grifoni;[37] Antonelli et al[38]	Q-panel UVB-313	HF	CA	PE and LNE	System description and vegetation	Manually modulated treatment for 7 h per day simulating 20% O_3 depletion at Pistoia, Italy, calculated using the model of Björn and Murphy.[24] *Phaseolus vulgaris* also under development.
N.D. Paul (pers. comm.)	Philips TL40/12	LF	CA	PE 'Mylar'	Fungal pathogens of crops	Timer controlled exposure to study wheat pathogen *Septoria tritici*.

[a]Reprinted with permission from: McLeod AR. Plant Ecol 1997; 28:78-92. © 1997 Kluwer Academic Publishers. [b]LF, low frequency (50/60 Hz); HF, high frequency; PE, polyester; CA, cellulose diacetate; LNE, lamps not energized.

Table 2.4. Modulated outdoor supplementation systems for studies of increased UV-B radiation[a,b]

Reference	Lamp type	High/low frequency	Filters Treatment	Filters Control	Sensor type	Species studied/experimental details
Caldwell et al;[39] Flint et al[40]	Westinghouse FS-40	HF	CA	PE 'Mylar-D'; no lamp control in *Vicia faba*	Filtered Hamamatsu UV photodiodes attached to master lamps	Dynamic range of modulation 50:1, R-B meter measured ambient UV-B; study of *Vicia faba* stimulating 6 and 32% O_3 depletion calculated using the model of Green et al.[27]
Caldwell et al[41]	Q-Panel UVB-313 and UVB-351	HF	Cellulose triacetate	PE	As above	*Glycine max*. Four experiments using different lamps and filter around the lamps and the lamp frames to vary the ratios of UV-B:UV-A:PAR. UV-B exposure simulating 36% O_3 depletion at Logan, U.S., calculated using the model of Green et al.[27]
Barnes et al;[42] Barnes et al[43]	Q-Panel UVB-313	HF	CA	PE	As above	Mixtures and monocultures of *Triticum aestivum* and *Avena fatua*. Dose equivalent to different O_3 depletions calculated for Logan, U.S., using the model of Green et al.[27]
Yu et al[44]	Q-Panel UVB-313	LF	CA	PE 'Mylar-S'	YMT, Interscience Tech, MD, U.S.	Dynamic range of modulation 300:1. Sensor response similar to plant action spectrum.[15]
Sullivan et al[29]	Q-Panel UVB-313	LF	CA	PE 'Mylar-S'	YMT, as above	Comparison of square-wave and modulated exposure of *Glycine max* at Beltsville, U.S.

Reference	Lamp			Lamp model	Description
Mepsted et al[45]	Philips TL40/12	LF	CA	LNE	BW-100, Vital Tech, Ontario, Canada, UV-B	Study of *Pisum sativum* at Wellesbourne, UK. Simulated 15% O_3 depletion calculated using model of Björn and Murphy[24] and following seasonal variations in the O_3 column. Sensor cross-calibrated to PAS.
Anonymous[46]	Philips TL40/12	LF	CA	LNE	BW-100, UV-B version as above	Field study of limestone grassland at Buxton, UK, using model of Björn and Murphy[24] as above to simulate 15, 20, 25 & 30% O_3 depletion and following seasonal variations in the O_3 column. Sensor cross-calibrated to PAS.
Campbell et al[47]	Philips TL40/12	HF	CA	LNE	BW-100, UV-B version as above	Simulation of 6 O_3 depletion levels (0-25%). Dynamic range of modulation 100:1.
N.D. Paul (pers. comm.)	Philips TL40/12	LF	CA	LNE	BW-100, UV-B version as above	Decomposition study of *Calluna vulgaris* and *Rubus chamaemorus* at Lancaster, UK. Simulated 15% O_3 depletion calculated using model of Björn and Murphy[24] and following seasonal variations in the O_3 column. Sensor cross-calibrated to PAS.
Coop et al[48]	Philips TL40/12	LF	CA	LNE	BW-100, UV-B version as above	Study at Lancaster, UK of *Calluna vulgaris* and *Vaccinium myrtillus* in Teflon tunnels to permit CO_2 interaction studies. Simulated 0, 15 and 25% O_3 depletion calculated using model of Björn and Murphy.[24]

Table 2.4. continued

Reference	Lamp type	High/low frequency	Filters Treatment	Filters Control	Sensor type	Species studied/experimental details
Nouchi and Kobayashi[49]	Philips F40UVB	LF	CA	PE 'Mylar-D'	MS-210D Eiko Seiki Japan	Rice cultivars. Modulated increase in ambient UV-B to 2.67 times measured ambient. Sensor response similar to DNA action spectrum.
Newsham et al[13]	Q-Panel UVB313	LF	CA	PE 'Mylar' and LNE	BW-100 CIE erythemal version	Modulated 30% increase above ambient UV-B. Study of *Hyacinthoides non-scripta*, *Anemone nemorosa* and *Quercus robur* at Monks Wood, UK.

[a]Reprinted with permission from: McLeod AR. Plant Ecol 1997; 28:78-92. © 1997 Kluwer Academic Publishers. [b]LF, low frequency (50/60 Hz); HF, high frequency; PE, polyester; CA, cellulose diacetate; LNE, lamps not energized.

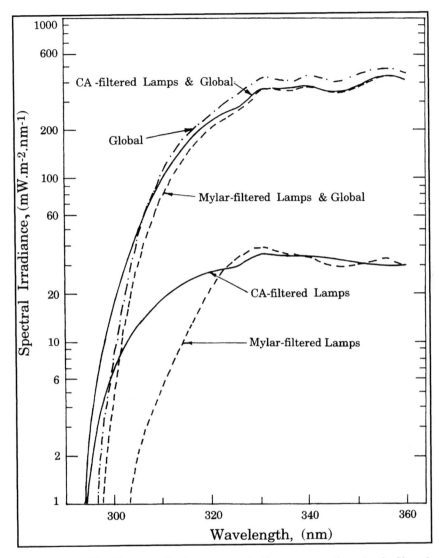

Fig. 2.3. Spectral irradiance received at 40 cm under fluorescent sunlamp banks filtered with plastic film filters. The lamps for UV-B enhancement (filtered with cellulose acetate, CA film) are adjusted to provide a UV-B supplement equivalent to a 16% ozone reduction under these conditions at 1200 solar time, on 20 August 1981, at 41°45'N latitude and 1460 m elevation. The control lamp banks (filtered with Mylar-D film) are adjusted to provide the same UV-A irradiance as is emitted by the CA-filtered lamps. Spectral irradiance from these lamp banks is shown with and without the background solar UV irradiance. Solar UV spectral irradiance above the lamp banks is also depicted. Reprinted with permission from: Caldwell MM, Gold WG, Harris G et al. Photochem Photobiol 1983; 37:479-485. © 1983 American Society for Photobiology.

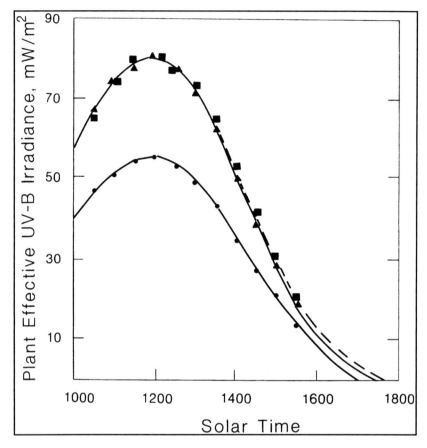

Fig. 2.4. Effective UV-B irradiance (using the generalized plant action spectrum weighting function) based on spectral irradiance measurements in the field for solar global irradiance above the lamp banks on 17 August 1981, in Logan, Utah (U.S.) and under the UV-enhancement lamp banks with the background solar irradiance. Theoretical effective UV-B solar irradiance with a 16% ozone layer reduction (from the Green et al[27] model) for these conditions is also shown. (●———●), solar irradiance; (■———■), actual enhancement; (▲- - -▲), expected enhancement. Reprinted with permission from: Caldwell MM, Gold WG, Harris G et al. Photochem Photobiol 1983; 37:479-485. © 1983 American Society for Photobiology.

The biologically effective UV-B dose or UV-B(BE) is calculated through the application of weighting functions derived from action spectra (e.g., damage to DNA, inhibition of photosynthesis) to assess the relative biological effectiveness of polychromatic irradiance:

$$\text{Effective irradiance } (Jm^{-2}) = \int I_\lambda E_\lambda d\lambda \quad (1)$$

where, I_λ = spectral irradiance ($Wm^{-2}s^{-1}nm^{-1}$), and E_λ = relative effectiveness of irradiance at wavelength λ to elicit a particular biological response. The limits of the integration are determined by the wavelength where either I_λ or E_λ approach zero.

According to Caldwell et al,[50] weighted effective irradiance addresses three basic issues:

1. The increment of biologically damaging irradiation resulting from a given level of O_3 reduction is known as the Radiation Amplification Factor (RAF).[51] RAF (Table 2.5) takes into account the biological effectiveness of each wavelength, however, the integrated measurements do not (Table 2.1). The increase in solar UV-B radiation as a function of O_3 reduction only becomes significant when the biological effectiveness of this radiation is calculated and the action spectrum has certain characteristics. If the action spectrum of a particular biological phenomenon does not exhibit appropriate characteristics, the RAF will be very small and the phenomenon under consideration may be eliminated from being one of concern;
2. At the earth's surface, as with O_3, there is a natural latitudinal and altitudinal gradient of solar UV-B irradiance which influences its biological effectiveness; and
3. Spectral irradiance received from commonly used lamp systems in artificial UV-B exposures does not match that of solar irradiance. Therefore, comparisons are only possible by calculating "biologically effective" radiation using action spectra as weighting functions. These three comments clearly summarize a frequently observed methodological problem at present.

Since action spectra differ greatly in their decline with increasing wavelength (Fig. 2.5), it is important to decide which of the primary injury mechanisms are the most important for intact higher plants exposed to solar radiation. To make this decision one needs to know: (a) the basic sensitivity of the respective targets (e.g., DNA, photosystem II) in the plant species under consideration; (b) the protection offered by the location of the target within the plant tissue and the actual effective UV-B fluence reaching it; the dose of interest is that which reaches the receptors, not the dose at the leaf surface; (c) the extent of reciprocity (i.e., that response is a function of the product or integral of irradiance and time); and (d) whether radiation at different wavelengths when applied simultaneously and

Table 2.5. Radiation amplification factors (RAFs) at 30°N[a,52]

Effect	RAF January	July	Reference
DNA related			
Mutagenicity and fibroblast killing	[1.7]2.2	[2.7]2.0	Zölzer and Kiefer,[53] Peak et al[54]
Fibroblast killing	0.3	0.6	Keyse et al[55]
Cyclobutane pyrimidine dimer formation	[2.0]2.4	[2.1]2.3	Chan et al[56]
(6-4) photoproduct formation	[2.3]2.7	[2.3]2.5	Chan et al[56]
Generalized DNA damage	1.9	1.9	Setlow[14]
HIV-1 activation	[0.1]4.4	[0.1]3.3	Stein et al[57]
Plant effects			
Generalized plant spectrum	2.0	1.6	Caldwell et al[50]
Inhibition of growth of cress seedlings	[3.6]3.8	3.0	Steinmetz and Wellman[58]
Isoflavonoid formation in bean	[0.1]2.7	[0.1]2.3	Wellman[59]
Inhibition of phytochrome-induced anthocyanin synthesis in mustard	1.5	1.4	Wellmann[59]
Anthocyanin formation in maize	0.2	0.2	Beggs and Wellmann[60]
Anthocyanin formation in sorghum	1.0	0.9	Yatsuhashi et al[61]
Photosynthetic electron transport	0.2	0.2	Bornman et al[62]
Overall photosynthesis in leaf of *Rumex patientia* L.	0.2	0.3	Rundel[63]
Membrane damage			
Glycine leakage from *Escherichia coli*	0.2	0.2	Sharma and Jagger[64]
Alanine leakage from *Escherichia coli*	0.4	0.4	Sharma and Jagger[64]
Membrane-bound K+-stimulated ATPase inactivation	[0.3]2.1	[0.3]1.6	Imbrie and Murphy[65]
Skin			
Elastosis	1.1	1.2	Kligman and Sayre[66]
Photocarcinogenesis, skin edema	1.6	1.5	Cole et al[67]
Erythema reference	1.1	1.1	McKinlay and Diffey[68]
Skin cancer in SKH-1 hairless mice	1.4	1.3	de Gruijl and van der Leun,[69] Longstreth et al[70]
Eyes			
Damage to cornea	1.2	1.1	Pitts et al[71]
Damage to lens (cataract)	0.8	0.7	Pitts et al[71]
Movement			
Inhibition of motility in *Euglena gracilis*	1.9	1.5	Häder and Worrest[72]

Table 2.5. continued

Effect	RAF		Reference
	January	July	
Other			
Immune suppression	[0.4]1.0	[0.4]0.8	De Fabo and Noonan[73]
Robertson-Berger meter	0.8	0.7	Urbach et al[74]

[a]Values in brackets show effect of extrapolating original data to 400 nm with an exponential tail, for cases where the effect is larger than 0.2 RAF units.

in proportion to their occurrence in sunlight would lead to an additive, more than additive or less than additive effect.[50]

Coohill[77] examined several UV action spectra that typify the responses of higher plants to irradiation by wavelengths between 280 and 380 nm. He concluded that no single action spectrum or solar effectiveness spectrum can be considered to be general enough to predict the varied responses of plants at wavelengths throughout the UV-B range. Caldwell and Flint[78] have discussed the issue of monochromatic versus polychromatic biological action spectra. They concluded that bottom-up mechanistic and top-down polychromatic action spectrum development are considered to be unsatisfactory. The authors offered an intermediate approach using whole plant morphological changes, but no supportive data.

There is a significant amount of variability in the procedures used for UV-B irradiance in artificial exposure experiments.[18,19,79,80] In addition, extrapolation of the spectra of component reactions of isolated organelles to the spectra of whole plant response under ambient conditions is highly precarious[50] and the application of such action spectra do not consider the role of polychromatic light and the phenomenon of photorepair. Coohill[77] correctly pointed out that, if common irradiation and response assay procedures are followed, then a general plant effect action spectrum may be forthcoming in the future. In contrast, Sutherland[81] concluded that although action spectra for UV-B radiation damage to higher organisms can give useful information, inappropriate analyses can lead to misleading and even incorrect results. According to the author it might be useful

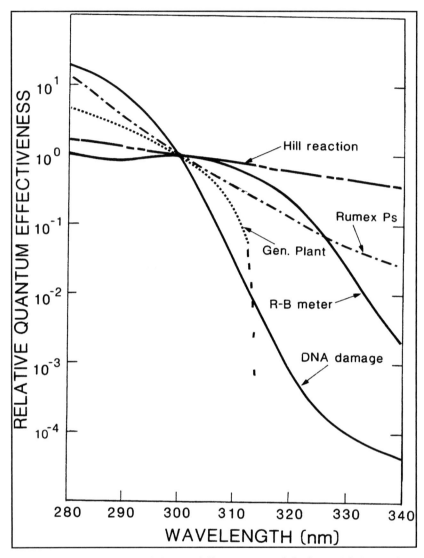

Fig. 2.5. Action spectra exhibiting different rates of decline with increasing wavelength. These include the spectrum for Hill reaction inhibition of spinach chloroplasts,[75] the *Rumex patientia* L. leaf photosynthesis inhibition spectrum, a spectrum for the Robertson-Berger meter,[76] a generalized plant damage spectrum[15] and a generalized DNA damage spectrum.[14] Reprinted with permission from: Caldwell MM, Camp LB, Warner CW et al. In: Worrest RC, Caldwell MM, eds. Stratospheric Ozone Reduction, Solar Ultraviolet Radiation and Plant Life. Berlin: Springer-Verlag, 1986:87-111. © 1986 Springer-Verlag.

to include both normalized and non-normalized action spectra in cases of complex biological responses.

A Comparative Analysis of O_3-CO_2-UV-B Dosimetry

An issue related to the use of single, seasonal or annual mean values to describe the stochastic exposure dynamics of many air pollutants such as O_3, is the concept of "dose"— a term frequently used in air pollution literature. A derivation from the concept of dose is the separation of the terms *"exposure dose"*[82] and *"effective dose."*[83] Exposure dose can be defined as the air concentration of the pollutant over a given duration to which the plant is subjected. To the contrary, effective dose can be defined as the actual pollutant concentration absorbed by the plant over a given duration. It is the effective dose that leads to a plant response. Runeckles[84] has discussed this concept more fully as it applies to O_3. While methodology is available for the quantification of effective ambient CO_2 dose,[85,86] similar techniques are not available for O_3 and UV-B. Rates of uptake or absorption of these two variables must be inferred through the computation of their flux to the plant canopy or receptor site.[79,87-89]

Ozone is known to be the most ubiquitous phytotoxic air pollutant worldwide.[6,90] In spite of this, present ambient O_3 monitoring is geographically sparse and inadequate, being mostly oriented to populated areas, rather than to agricultural and forested areas. Nevertheless, this database is much more extensive than similar efforts to monitor ambient CO_2 and UV-B. Over the past two decades numerous investigators have studied and modeled vegetation response to chronic O_3 exposure.[82,91-93] There are a number of issues relevant to these efforts. A discussion of some of these issues can be useful in future research on the joint effects of elevated O_3, CO_2 and UV-B on vegetation.

One issue concerns the numerical definition of ambient O_3 exposure kinetics to which vegetation is subjected. In most cases annual or seasonal mean or summation of O_3 concentrations of various types have been used in modeling cause-effect relationships.[91-93] The motivation behind these approaches has been to find a simple exposure statistic that will provide information for establishing ambient regulatory air quality standards, objectives or guidelines for O_3. However, presently this is an area of much controversy.[94] Growth season hourly ambient O_3 concentrations exhibit a non-normal frequency distribution. In these situations, computations of seasonal or annual means or summations are statistically inappropriate.[93,95] This point was previously confirmed by Legge et al[96] who showed decreasing values of the coefficient of determination (R^2) between the seasonal mean O_3 concentrations and increasing values of the seasonal hourly maximum O_3 concentrations at multiple O_3 monitoring sites and years.

In consideration of the aforementioned concerns and similar others, some investigators have applied weighting functions to describe the O_3 exposure. One approach consists of differentially weighting different ambient O_3 concentrations, with the high O_3 concentrations given the most weight.[97] This is based on the concept that progressively higher air concentrations of O_3 have progressively higher potential to cause stress to vegetation. Recently, this concept was examined by Grünhage et al[89] who were unable to establish a consistent relationship between high air concentrations of O_3, its flux onto a grassland canopy and plant uptake. Furthermore, these types of weighting functions are unable to account for the stochastic relationships between the varying time series of ambient O_3 and the dynamics of plant growth. There is evidence to show that different phenological or growth stages of a given plant species respond differently to O_3 and sulfur dioxide stress.[97,98] To address this issue in the context of O_3, at least one study has attempted the use of a phenological weighting function in examining plant response.[99] However, in this case the effort was strictly a computer based statistical analysis and not validated by actual experimental data.

The response of certain plant species to elevated CO_2 is somewhat analogous to the aforementioned comments for O_3. According to Krupa and Kickert,[12] sorghum (*Sorghum vulgare* (L.) Moench) (C4) showed almost no response to elevated CO_2 during its early stages compared to the whole season. In contrast, cotton (*Gossypium hirsutum* L.) (C3) was very responsive in both the early stages of its growth and over the entire season. Cucumber (*Cucumis sativus* L.) (C3) and radish (*Raphanus sativus* L.) (C3), while very responsive in their early stages of growth, showed a progressive decrease in response over the whole growth season.[12] Similar temporal variability in plant response to UV-B is expected, since leaf properties are known to vary with plant age and growth environment.[100]

To address some of these considerations, Krupa and Nosal[101] in their studies on ambient O_3 and alfalfa (*Medicago sativa* L.) response, applied spectral coherence analysis to account for weekly exposure dynamics of O_3 and the corresponding changes in height growth of the plants. After testing a number of numerical definitions of the O_3 exposure parameters, the aforementioned authors concluded that during non-episodal (respite) periods, the median value of the hourly O_3 concentrations for those periods (if occurring during the proper diurnal period), was in best coherence with the corresponding changes in alfalfa height growth. In contrast, during episodal (stress) periods, the cumulative integral of the hourly O_3 concentrations for those periods was in best coherence with the corresponding changes in alfalfa height growth. While this approach is strictly statistical in nature, it allows for an improved numerical definition of the stochastic independent variable(s). Having done so, such definitions must be incorporated into a corresponding time series-based predictive biological process model to achieve wide applicability and reproducibility.[93] While such efforts most likely can account for the temporal and spatial variability in cause-effect relationships, they would require extensive and intensive research efforts before the concept could be considered in regulatory policies. Nevertheless, it deserves scientific merit.

Krupa et al[102] performed a retrospective numerical analysis of the O_3 exposure-crop response data from the U.S. NCLAN (National Crop Loss Assessment Network). Because of the nature of the data, only the relationships between O_3 exposures and season-end crop harvests could be examined (single-point model). Such an analysis did not allow for an inclusion of feedback from the differing responses of different crop growth stages to O_3 exposures (multi-point model). An understanding of the feedback is necessary to develop generalized, reproducible O_3 dose-crop response relationships. Nevertheless, Krupa et al[102] found that the cumulative frequency of the occurrence of hourly O_3 concentrations > 49 ppb was much higher for cases where O_3 exposure resulted in crop yield reductions, in comparison to the corresponding frequency for cases with no yield reductions. Further, this study showed that the cumulative frequency of hourly O_3 concentrations between 50 ppb and 87 ppb was the best predictor of the crop yield reductions ($R^2 = 0.76$, $p = 0.011$). This is very impressive, since the technique used was a simple, univariate regression. These results were further expanded and verified using other independent data sets on O_3 exposure and crop response (e.g., data from the European Open-top Chambers Programme) by Legge et al[103] and Grünhage et al.[89]

A second controversial and difficult issue in O_3 effects research concerns what a control should be in experimental field studies. A number of investigators (with the exception of Krupa et al,[102] have used a seasonal mean O_3 concentration of 0.025 ppm as the control in modeling cause-effect relationships.[92] Independent of the discussion on the inappropriateness of using seasonal mean values of non-normally distributed data, Heuss,[104] Lefohn et al[105,106] and Legge et al[96] have questioned the use of a 0.025 ppm constant mean O_3 value as the control. According to these authors the so-called background O_3 concentrations will vary in time and space, possibly from 0.025 to 0.045 ppm. Given this situation, use of a constant, mean O_3 concentration of 0.025 ppm in any regression model will result in an overestimation of any adverse effects on vegetation.[105] This is a very valid argument since at a given geographic site there will be inter- and intra-annual variability in the ambient O_3 levels. Given this fact that atmospheric O_3 production and decomposition are highly variable, it is unrealistic to consider a constant control from season-to-season and location-to-location.[96] If effects scientists accept response surface methodology[107] more readily, then the range of the true control (intercept in a regression) can more readily be derived mathematically and compared against the experimental control. This, in fact would be an important scientific contribution.

In contrast to the aforementioned concerns regarding O_3, many CO_2 exposure experiments under field conditions have addressed the issue of elevated CO_2 concentrations beyond the present,[12] because of the increasing concentrations of that pollutant.[108] In terms of science, one should also include in such experiments, a treatment or treatments with CO_2 concentrations of the past, for example, 1940s or 1950s.[109] This would be particularly useful in the context of studies on joint effects of multiple independent variables. This is also true in the context of UV-B studies, where much concern has been directed to its future elevated levels. At least two studies,[4,110] have shown a decrease in the surface UV-B over the past several decades in the U.S., perhaps due to increases in surface O_3 and fine particle concentrations. Independent of this, we really do not have experimental data on the joint effects of CO_2, UV-B, O_3, temperature and moisture, regardless of how they vary at the present time and in the future.

Altered Ambient CO_2, UV-B, O_3, Temperature and Moisture Levels

Many of us tend to view plant health in the ambient environment as a univariate system, as we have, for example, in many of our studies on the potential adverse effects of acidic precipitation on terrestrial vegetation.[111] In many of our field experiments, what is expected to serve as a control for a given independent variable, most likely does not serve as a realistic control for another independent variable of concern. Given what we know, greater emphasis should be directed to experiments conducted in open, ambient environments so as to increase our confidence in the results we obtain. While traditionally effects scientists are concerned about the complexity attached to this approach, systems level modelers do not have a similar concern.[93] Equally important is the fact that there are few studies where atmospheric scientists, mathematicians and biologists have truly joined together to conduct a specific study in common time and space, from start to finish on a continuous basis.[112]

Figure 2.6 shows potential simultaneous and sequential exposure dynamics of multiple stress factors of concern relevant to future vegetation (crop) response research. Such exposure dynamics, according to Krupa and Kickert,[12] can result, as an example, in:

1. No interaction between the stress factors. The "Law of Limiting Factors" might prevail in which the most severe stress overrides plant response; or
2. A cumulative effect in which the net plant response is simply the sum of stress effects (additive) from O_3 and increased UV-B regardless of the temporal patterns of exposure; or
3. There might be a more than additive effect where the plant response is more severe than would be found from either stress singly. There is also the possibility of a less than additive interaction in the sense that high ambient CO_2 might allow sufficient repair processes to proceed in some plants so that sensitivity to increased UV-B and/or ambient O_3 may be reduced.

If one conceives of mathematical functions, or graphs, where the "UV-B effect" and "ozone effect" on net photosynthesis (P_{NA}) as an example are scaled between 0.0 to 1.0 as functions of UV-B(BE) and ambient O_3 exposure respectively, then, as a first approximation

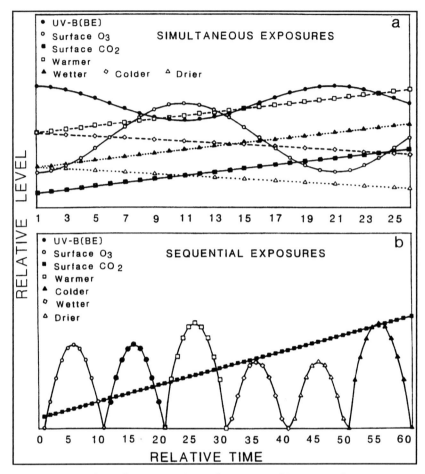

Fig. 2.6. A hypothetical scenario (a) simultaneous vs. (b) sequential exposures to potential changes in multiple climatic variables. Reprinted with permission from: Krupa SV, Kickert RN. Vegetatio 1993; 104/105:223-238. © 1993 Kluwer Academic Publishers.

to mathematical model development, we consider the following counterparts to the three hypotheses stated above:

1. $P_{NA} = P_{NCO_2} * AMIN$ (UV-B effect, O_3 effect)

where P_{NCO_2} is the net photosynthesis modeled as a response to increased CO_2, P_{NA} is net photosynthesis after adjustment for UV-B and/or O_3, and AMIN is a computer program function that means "use the minimum value of the variables in parenthesis" which actually represent the most severe stress;

2. $P_{NA} = P_{NCO_2} * (1 - \text{AMIN} [(1 - \text{UV-B effect}) + (1 - O_3 \text{ effect}), 1])$

3. $P_{NA} = P_{NCO_2} * C * (\text{UV-B effect} * O_3 \text{ effect})$

where C is a coefficient of proportionality. This set of alternative hypotheses could be imbedded within a larger, comprehensive crop growth model run day-by-day over the growth season for the purpose of conducting computer simulation experiments.

For those plant species that show sensitivity to any two of the environmental stimuli, O_3, enhanced UV-B radiation, or increased CO_2, or especially for those species that are sensitive to all of these stimuli, serious questions must be raised about the results of ambient field exposures of such plants to either O_3, enhanced UV-B or increased CO_2 alone. One of the main effects of O_3 is on photosynthesis. We know of no ambient field exposures of plants to O_3 in which the study also included measurements of natural UV-B and ambient CO_2 concentrations or fluxes. Any plant effects not attributable to O_3, which might have occurred in such studies would be unidentified and masked in the error terms of any quantitative analyses. Likewise, none of the open field experiments of enhanced UV-B radiation on plants have included the measurements of ambient O_3 or any other air pollutant. Accordingly, results of such studies could be confounded by the effects of pollutants such as the increase in CO_2, in addition to failing to describe microclimatic flows of radiant and heat energy and moisture, for comparison to analogous studies in artificial exposure environments. However, if any increase in CO_2 in the field is a very stable long-term process without a high frequency of variability, it simply means that the relative level of effects between plant species and cultivars under experimentally enhanced UV-B radiation might not be affected by the long-term increase in CO_2. The absolute level of effects would, however, be unknown because we do not know the past concentrations of CO_2 at a given geographic environment to which plant species and cultivars have become adapted over time.

One way out of this dilemma in the future is for field experiments to include monitoring and analysis of all three potential growth influencing factors, in addition to the more common considerations generally given to soil and meteorological constraints, as well as the effects of biotic pathogens and pests on plant growth. First order numerical time series models which can accommodate such measurements in evaluating cause-effects relationships are presently available.[113,114] However, such models must be integrated with approaches to plant pest and disease epidemiology and would require the use of main-frame computers and innovative mathematical concepts.

References

1. Webb AR. Solar ultraviolet-B radiation measurement at the earth's surface: Techniques and trends. In: Abrol YP, Govindjee, Wattal PN et al, eds. Impact of Global Climatic Changes on Photosynthesis and Plant Productivity. New Delhi: Oxford & Ibh Publ. Co. PVT. Ltd., 1991:23-37.
2. German Bundestag. Ozone depletion, changes in UV-B radiation and their effects. In: Protecting the Earth—A Status Report With Recommendations for a New Energy Policy. Bonn: Deutscher Bundestag, Referat Offentlichkeirsarbeit, 1991:571-616.
3. Cutchis P. Stratospheric ozone depletion and solar ultraviolet radiation on earth. Science 1974; 184:13-19.
4. Frederick JE, Snell HE, Haywood EK. Solar ultraviolet radiation at the earth's surface. Photochem Photobiol 1989; 50:443-450.
5. Runeckles VC, Krupa SV. The impact of UV-B and ozone on terrestrial vegetation. Environ Pollut 1994; 83:191-213.
6. Krupa SV, Manning WJ. Atmospheric ozone: Formation and effects on vegetation. Environ Pollut 1988; 50:101-137.
7. Chevone BI, Herzfeld DE, Krupa SV et al. Direct effects of atmospheric sulfate deposition on vegetation. JAPCA 1986; 36:813-815.
8. U.S. NAPAP. Acidic Deposition: State of Science and Technology. Summary Report. Washington, DC: U.S. National Acid Precipitation Assessment Program, 1991.
9. Krupa SV, Kickert RN. The Greenhouse Effect: The impacts of carbon dioxide (CO2), ultraviolet-B (UV-B) radiation and ozone (O_3) on vegetation (crops). Vegetatio 1993; 104/105:223-238.
10. Gardiner BG, Webb AR, Bais AF et al. European intercomparison of ultraviolet spectroradiometers. Environ Technol 1993; 14:25-43.
11. Bernhard G, Seckmeyer G. New entrance optics for solar spectral UV measurements. Photochem Photobiol 1997; 65:923-930.
12. Krupa SV, Kickert RN. The Greenhouse Effect: Impacts of ultraviolet-B (UV-B) radiation, carbon dioxide (CO_2), and ozone (O_3) on vegetation. Environ Pollut 1989; 61(4):263-393.
13. Newsham KK, McLeod AR, Greenslade PD et al. Appropriate controls in outdoor UV-B supplementation experiments. Global Change Biol 1996; 2:319-324.
14. Setlow RB. The wavelengths in sunlight effective in producing skin cancer: A theoretical analysis. Proc Natl Acad Sci USA 1974; 71:3363-3366.
15. Caldwell MM. Solar UV irradiation and the growth and development of higher plants. In: Giese AC, ed. Photophysiology, Vol. VI,

Current Topics in Photobiology and Photochemistry. New York: Academic Press, 1971:131-177.
16. Coohill TP. Stratospheric ozone depletion as it affects life on earth—The role of ultraviolet action spectroscopy. In: Abrol YP, Govindjee, Wattal PN et al, eds. Impact of Global Climatic Changes on Photosynthesis and Plant Productivity. New Delhi: Oxford & Ibh Publ. Co. PVT. Ltd., 1991:3-21.
17. Iwanzik W. 1986. Interaction of UV-A, UV-B and visible radiation on growth, composition and photosynthetic activity in radish seedlings. In: Worrest RC, Caldwell MM, eds. Stratospheric Ozone Reduction, Solar Ultraviolet Radiation and Plant Life. Berlin: Springer-Verlag, 1986:287-301.
18. Tevini M, Mark U, Saile M. Plant experiments in growth chambers illuminated with natural sunlight. In: Payer HD, Pfirrmann T, Mathy P, eds. Environmental Research With Plants in Closed Chambers. Air Pollution Research Report No. 26. Brussels: Commission of the European Communities, 1990:240-251.
19. Krizek DT, Mirecki RM, Philbeck RB et al. An improved lamp/filter system for use in UV enhancement studies under greenhouse and field conditions. In: Payer HD, Pfirrmann T, Mathy P, eds. Environmental Research With Plants in Closed Chambers. Air Pollution Research Report No. 26. Brussels: Commission of the European Communities, 1990:271-278.
20. Steeneken S, Buma AGJ, Gieskes WWC. Changes in transmission characteristics of polymethylmethacrylate and cellulose (III) acetate during exposure to ultraviolet light. Photochem Photobiol 1995; 61:276-280.
21. Johanson U, Gehrke C, Björn LO et al. The effects of enhanced UV-B radiation on a subarctic heath ecosystem. Ambio 1995; 24:106-111.
22. Johanson U, Gehrke C, Björn LO et al. The effects of enhanced UV-B radiation on the growth of dwarf shrubs in a subarctic heathland. Funct Ecol 1995; 9:713-719.
23. Gehrke C, Johanson U, Callaghan TV et al. The impact of enhanced ultraviolet-B radiation on litter quality and decomposition processes in *Vaccinium* leaves from the subarctic. Oikos 1995; 72:213-222.
24. Björn LO, Murphy TM. Computer calculation of solar ultraviolet radiation at ground level. Physiol Veg 1985; 23:555-561.
25. Gehrke C, Johanson U, Gwynn-Jones D et al. Single and interactive effects of enhanced ultraviolet-B radiation and increased atmospheric CO_2 on terrestrial subarctic ecosystems. Ecol Bull 1996; 45:192-203.
26. Teramura AH, Sullivan JH, Lyden J. Effects of UV-B radiation on soybean yield and seed quality: A 6-year field study. Physiol Plant 1990; 80:5-11.
27. Green AES, Cross KR, Smith LA. Improved analytical characterization of ultraviolet skylight. Photochem Photobiol 1980; 31:59-65.
28. Sullivan JH, Teramura AH. The effects of ultraviolet-B radiation on loblolly pine. 2. Growth of field-grown seedlings. Trees Struct Funct 1992; 6:115-120.

29. Sullivan JH, Teramura AH, Adamse P et al. Comparison of the response of soybean to supplemental UV-B radiation supplied by either square-wave or modulated irradiation systems. In: Biggs RH, Joyner MEB, eds. Stratospheric Ozone Depletion/UV-B Radiation in the Biosphere. NATO ASI Series. Berlin: Springer-Verlag, 1994:211-220.
30. Booker FL, Fiscus EL, Philbeck RB et al. A supplemental ultraviolet-B radiation system for open-top field chambers. J Environ Qual 1992; 21:56-61.
31. Fiscus EL, Miller JE, Booker FL. Is UV-B a hazard to soybean photosynthesis and yield? Results of an ozone UV-B interaction study and model predictions. In: Biggs RH, Joyner MEB, eds. Stratospheric Ozone Depletion/UV-B Radiation in the Biosphere. NATO ASI Series. Berlin: Springer-Verlag, 1994:135-147.
32. Petropoulou Y, Kyparissis A, Nikolopoulos D et al. Enhanced UV-B radiation alleviates the adverse effects of summer drought in two Mediterranean pines under field conditions. Physiol Plant 1995; 94:37-44.
33. Zeuthen J. UV-B stralingens indflydelsepa egemeldung (The effect of UV-B radiation on oak mildew). Skoven 1995; 12:490-493.
34. Dai Q, Peng S, Chavez AQ et al. Effect of enhanced UV-B radiation on growth and production of rice under greenhouse and field conditions. In: Peng S, Ingram KT, Neue H-U et al, eds. Climate Change and Rice. Berlin: Springer-Verlag, 1995:242-257.
35. Olszyk D, Dai Q, Teng P et al. UV-B effects on crops: Response of the irrigated rice ecosystem. J Plant Physiol 1996; 148:26-34.
36. Rozema J, Tosserams M, Magendans E. Impact of UV-B radiation on plants from terrestrial ecosystems. In: Zwerver S, Rompaey V, Kok MTJ et al, eds. Climate Change Research: Evaluation and Policy Implications. Amsterdam: Elsevier Science, 1995:997-1004.
37. Zipoli G, Grifoni D. Effects of UV-B radiation on vegetation: A field apparatus for artificial UV-B supplementation. Proc. 3rd ESA Congress, Abano-Padova. 1995:424-425.
38. Antonelli F, Grifoni D, Sabatini F et al. Morphological and physiological responses of bean plants to supplemental UV radiation in a Mediterranean climate. Vegetatio 1996; 128:127-136.
39. Caldwell MM, Gold WG, Harris G et al. A modulated lamp system for solar UV-B (280-320 nm) supplementation studies in the field. Photochem Photobiol 1983; 37:479-485.
40. Flint SD, Jordan PW, Caldwell MM. Plant protective response to enhanced UV-B radiation under field conditions: Leaf optical properties and photosynthesis. Photochem Photobiol 1985; 41:95-99.
41. Caldwell MM, Flint SD, Searles PS. Spectral balance and UV-B sensitivity of soybean: A field experiment. Plant Cell Environ 1994; 17:267-276.
42. Barnes JD, Paul ND, Percy K et al. Effects of UV-B radiation on wax biosynthesis. In: Percy K, Cape JN, Jagels R, eds. Air Pollutants

and the Leaf Cuticle. NATO-ASI Series. Berlin: Springer-Verlag, 1994:195-204.
43. Barnes PW, Flint SD, Caldwell MM. Early-season effects of supplemented solar UV-B radiation on seedling emergency, canopy structure, simulated stand photosynthesis and competition for light. Glob Change Biol 1985; 1:43-53.
44. Yu W, Sullivan JH, Teramura AH. The YMT ultraviolet-B irradiation system: Manual of operation. Final Report. Corvallis, OR: U.S. EPA Environ. Res. Lab., 1991.
45. Mepsted R, Paul ND, Stephen J et al. Effects of enhanced UV-B radiation on pea (*Pisum sativum* L.) grown under field conditions in the United Kingdom. Glob Change Biol 1996; 2:325-334.
46. Anonymous. Annual Report of the Unit of Comparative Plant Ecology. UK: Natural Environment Research Council, 1995.
47. Campbell BD, Pointon RF, Hunt CL. Progress in developing a technique for field-based UV-B enrichment. AgResearch Internal Report. Palmerston North, New Zealand: Grasslands Research Centre, 1995.
48. Coop DJS, Moody SM, Paul ND et al. Interactive effects of enhanced UV-B and CO_2 on northern heathlands. In: Proc. First European Symposium on the Effects of Environmental UV-B Radiation on Health and Ecosystems, 27-29 October 1993, Munich. European Commission DG-XIII, 1995:257-261.
49. Nouchi I, Kobayashi K. Effects of enhanced ultraviolet-B radiation with a modulated lamp control system on growth of 17 rice cultivars in the field. J Agr Meteorol 1995; 51:11-20.
50. Caldwell MM, Camp LB, Warner CW et al. Action spectra and their key role in assessing biological consequences of solar UV-B radiation change. In: Worrest RC, Caldwell MM, eds. Stratospheric Ozone Reduction, Solar Ultraviolet Radiation and Plant Life. Berlin: Springer-Verlag, 1986:87-111.
51. National Academy of Sciences. Protection Against Depletion of Stratospheric Ozone by Chlorofluorocarbons. Washington, DC: National Academy Press, 1979.
52. UNEP. Environmental Effects of Ozone Depletion: 1991 Update. Van der Leun JC, Tevini M, Worrest WC, eds. Nairobi, Kenya: United Nations Environmental Programme, 1991.
53. Zölzer F, Kiefer J. Wavelength dependence of inactivation and mutation induction to 6-thioguanine resistance in V79 Chinese hamster fibroblasts. Photochem Photobiol 1984; 40:49-53.
54. Peak MJ, Peak JG, Mohering MP et al. Ultraviolet action spectra for DNA dimer induction, lethality and mutagenesis in *Escherichia coli* with emphasis on the UVB region. Photochem Photobiol 1984; 40:613-620.
55. Keyse SM, Moses SH, Davies DJG. Action spectra for inactivation of normal and xeroderma pigmentosum human skin fibroblasts by ultraviolet radiation. Photochem Photobiol 1983; 37:307-312.
56. Chan GL, Peak MJ, Peak JG et al. Action spectrum for the formation of endonuclease-sensitive sites and (6-4) photoproducts induced

in a DNA fragment by ultraviolet radiation. Int J Radiat Biol 1986; 50:641-648.
57. Stein B, Rahmsdorf HJ, Steffen A et al. UV-induced DNA damage is an intermediate step in UV-induced expression of Human Immunodeficiency Virus Type 1, collagenase, C-fos and metallothionein. Mol Cell Biol 1989; 9:5169-5181.
58. Steinmetz V, Wellman E. The role of solar UV-B in growth regulation of cress (*Lepidium sativum* L.) seedlings. Photochem Photobiol 1986; 43:189-193.
59. Wellman E. UV-B signal/response-beziehungen unter natürlichen und artifiziellen lichtbedingungen. Ber Deutsch Bot Ges 1985; 98:99-104.
60. Beggs CJ, Wellmann E. Analysis of light-controlled anthocyanin formation in coleoptiles of *Zea mays*. 1. The role of UV-B, blue, red and far-red light. Photochem Photobiol 1985; 41:481-486.
61. Yatsuhashi H, Hashimoto T, Shimizu S. Ultraviolet action spectrum for anthocyanin formation in broom sorghum first internodes. Plant Physiol 1982; 70:735-741.
62. Bornman JF, Björn LO, Akerlund H-E. Action spectrum for inhibition by ultraviolet radiation of Photosystem II activity in spinach thylakoids. Photobiochem Photobiophys 1984; 8:305-313.
63. Rundel RD. Action spectra and the estimation of biologically effective UV radiation. Physiol Plant 1983; 58:360-366.
64. Sharma RC, Jagger J. Ultraviolet (254-405 nm) action spectrum and kinetic studies of analine uptake in *Escherichia coli* B/R. Photochem Photobiol 1979; 30:661-666.
65. Imbrie CW, Murphy TM. UV-action spectrum (254-405 nm) for inhibition of a K^+-stimulated adenosine triphosphatase from a plasma membrane of *Rosa damascena*. Photochem Photobiol 1982; 36:537-542.
66. Kligman LH, Sayre RM. An action spectrum for ultraviolet-induced elastosis in hairless mice: Quantification of elastosis by image analysis. Photochem Photobiol 1991; 53:237-242.
67. Cole CA, Forbes D, Davies RE. An action spectrum for UV photocarcinogenesis. Photochem Photobiol 1986; 43:275-284.
68. McKinlay AF, Diffey BL. A reference action spectrum for ultraviolet-induced erythema in human skin. In: Passchler WR, Bosnajokovic BFM, eds. Human Exposure to Ultraviolet Radiation: Risks and Regulations. Amsterdam: Elsevier, 1987.
69. de Gruijl FR, van der Leun JC. Action spectra for carcinogenesis contribution. Symposium on the Biologic Effects of UV-A Radiation, San Antonio, TX. 1991.
70. Longstreth JD, de Gruijl FR, Takizawa Y et al. Health effects. In: van der Leun JC, Tevini M, Worrest RC, eds. Environmental Effects of Ozone Depletion: 1991 Update. Nairobi, Kenya: UNEP, 1991.
71. Pitts DG, Cullen AP, Hacker PD. Ocular effects of ultraviolet radiation from 295 to 365 nm. Invest Ophthalmol Visual Sci 1977; 16:932-939.

72. Häder D-P, Worrest RC. Effects of enhanced solar ultraviolet radiation on aquatic ecosystems. Photochem Photobiol 1991; 53:717-725.
73. De Fabo EC, Noonan FP. Mechanism of immune suppression by ultraviolet radiation in vivo. I. Evidence for the existence of a unique photoreceptor in skin and its role in photoimmunology. J Exp Med 1983; 158:84-98.
74. Urbach F, Berger D, Davies RE. Field measurements of biologically effective UV radiation and its relation to skin cancer in man. In: Broderick AJ, Hard TM, eds. Proceedings of the Third Conference on Climatic Impact Assessment Program. Washington, DC: U.S. Department of Transportation. 1974.
75. Jones LW, Kok B. Photoinhibition of chloroplast reactions. I. Kinetics and action spectra. Plant Physiol 1966; 41:1037-1043.
76. Robertson DF. The sunburn unit for comparison of variation of erythemal effectiveness. In: Nachtwey DS, Caldwell MM, Biggs RH, eds. Impacts of Climatic Change on the Biosphere, Part I, Ultraviolet Radiation Effects, Monograph 5. Climatic Impact Assessment Program, U.S. Dept. Transportation, Report No. DOT-TST-75-55. Springfield, VA: NTIS, 1975.
77. Coohill TP. Ultraviolet action spectra (280-380 nm) and solar effectiveness spectra for higher plants. Photochem Photobiol 1989; 50(4):451-458.
78. Caldwell MM, Flint SD. Uses of biological spectral weighting functions and the need of scaling for the ozone reduction problem. Plant Ecol 1997; 128:66-76.
79. Worrest RC, Caldwell MM, eds. Stratospheric Ozone Reduction, Solar Ultraviolet Radiation and Plant Life. Berlin: Springer-Verlag, 1986.
80. Caldwell MM, Flint SD. Plant response to UV-B radiation: Comparing growth chamber and field environments. In: Payer HD, Pfirrmann T, Mathy P, eds. Environmental Research With Plants in Closed Chambers. Air Pollution Research Report No. 26. Brussels: Commission of the European Communities, 1990:264-270.
81. Sutherland BM. Action spectroscopy in complex organisms: Potentials and pitfalls in predicting the impact of increased environmental UVB. Photochem Photobiol 1995; 31:29-34.
82. Krupa SV, Kickert RN. An analysis of numerical models of air pollutant exposure and vegetation response. Environ Pollut 1987; 44(2):127-158.
83. Runeckles VC. Dosage of air pollutants and damage to vegetation. Environ Conserv 1974; 1:305-308.
84. Runeckles VC. Uptake of ozone by vegetation. In: Lefohn AS, ed. Surface Level Ozone Exposures and Their Effects on Vegetation. Chelsea, MI: Lewis Publishers Inc., 1992:157-188.
85. Enoch HZ, Kimball BA, eds. Carbon Dioxide Enrichment of Greenhouse Crops. II. Physiology, Yield, and Economics. Boca Raton: CRC Press, 1986.

86. Strain BR, Cure JD, eds. Direct Effects of Increasing Carbon Dioxide on Vegetation. DOE/ER-0238. Washington, DC: Office of Energy Research, U.S. Dept. of Energy, 1985.
87. Grünhage L, Jäger H-J. Influence of the atmospheric conductivity on the ozone exposure of plants under ambient conditions: Considerations for establishing ozone standards to protect vegetation. Environ Pollut 1994; 85:125-129.
88. Grünhage L, Dämmgen U, Haenel H-D et al. A new potential air quality criterion derived from vertical flux densities of ozone and from plant response. Angew Bot 1993; 67(1/2):9-13.
89. Grünhage L, Jäger H-J, Haenel H-D et al. PLATIN (PLant-ATmosphere INteraction): II. Co-occurrence of high ambient ozone concentrations and factors limiting plant absorbed dose. Environ Pollut 1997; (in press).
90. Kickert RN, Krupa SV. Forest responses to tropospheric ozone and global climate change: An analysis. Environ Pollut 1990; 68:29-65.
91. Environmental Pollution. Response of crops to air pollutants. Environ Pollut 1988; 53(1-4):1-478.
92. Heck WW, Tingey DT, Taylor OC, eds. Assessment of Crop Loss From Air Pollutants. London: Elsevier Applied Science, 1988.
93. Kickert RN, Krupa SV. Modeling plant response to tropospheric ozone: A critical review. Environ Pollut 1991; 70:271-383.
94. U.S. EPA. Workshop on Research Needs to Assess the Effects of Ozone on Crop, Forest and Natural Ecosystems. Draft Report. Heck WW, Furiness C, Sims C et al, eds. Raleigh, NC: The Southern Oxidants Study, North Carolina State University, 1997.
95. Krupa SV, Kickert RN. Considerations for establishing relationships between ambient ozone (O_3) and adverse crop response. Environ Rev 1997; 5:55-77.
96. Legge AH, Nosal M, McVehil GE et al. Ozone and the clean troposphere: Ecological implications. Environ Pollut 1991; 70:157-175.
97. Lefohn AS, Runeckles VC. Establishing a standard to protect vegetation-ozone exposure/dose considerations. Atmos Environ 1987; 21:561-568.
98. van Haut H, Stratmann H. Farbtafelatlas Über Schwefeldioxidwirkungen und Pflanzen. Essen: Verlag W. Girardet, 1970.
99. Lee EH, Tingey DT, Hogsett WE. Evaluation of ozone exposure indices in exposure-response modeling. Environ Pollut 1988; 53:43-62.
100. Martin JT, Juniper BE. The Cuticles of Plants. New York: St. Martins, 1970.
101. Krupa SV, Nosal M. Application of spectral coherence analysis to describe the relationships between ozone exposure and crop growth. Environ Pollut 1989; 60:319-330.
102. Krupa SV, Nosal M, Legge AH. Ambient ozone and crop loss: Establishing a cause-effect relationship. Environ Pollut 1994; 83:269-276.
103. Legge AH, Grünhage L, Nosal M et al. Ambient ozone and adverse crop response: An evaluation of North American and European data

as they relate to exposure indices and critical levels. Angew Bot 1995; 69: 192-205.
104. Heuss JM. Comment on assessment of crop loss from ozone. JAPCA 1982; 32(11):1152-1153.
105. Lefohn AS, Runeckles VC, Krupa SV et al. Important considerations for establishing a secondary ozone standard to protect vegetation. JAPCA 1989; 39(8):1039-1045.
106. Lefohn AS, Krupa SV, Winstanley D. Surface ozone exposures measured at clean locations around the world. Environ Pollut 1990; 63:189-224.
107. Myers RH. Response Surface Methodology. Boston: Allyn and Bacon, 1971.
108. EarthQuest. Boulder, CO: University Corporation for Atmospheric Research, 1990; 4(2):1-20.
109. Johnson HB, Wayne Polley H, Mayeux HS. 1993. Increasing CO_2 and plant-plant interactions: Effects on natural vegetation. In: Rozema J, Lambers H, van de Geijn SC et al, eds. CO_2 and Biosphere. Dordrecht: Kluwer Academic Publishers 1993:157-172.
110. Scotto J, Cotton G, Urbach F et al. Biologically effective ultraviolet radiation: Surface measurements in the United States, 1974 to 1985. Science 1988; 239:762-764.
111. Lefohn AS, Krupa SV, eds. Acidic precipitation: A technical amplification of NAPAP's findings. In: Proceedings of an APCA International Conference. Pittsburgh, PA: APCA, 1988.
112. Legge AH, Krupa SV, eds. Acidic Deposition: Sulphur and Nitrogen Oxides. Chelsea, MI: Lewis Publishers, Inc., 1990.
113. Krupa SV, Nosal M. Effects of ozone on agricultural crops. In: Schneider T, Lee SD, Wolters GJR, et al, eds. Atmospheric Ozone Research and Its Policy Implications. Amsterdam: Elsevier Science Publ. B.V., 1989:229-238.
114. Krupa SV, Nosal M. A multivariate, time series model to relate alfalfa responses to chronic, ambient sulfur dioxide exposures. Environ Pollut 1989; 61:3-10.

CHAPTER 3

Elevated UV-B Radiation and Crops

Introduction

Solar UV-B radiation as a portion of the electromagnetic radiant energy spectrum is often characterized by its wavelength. It has become an accepted practice to consider UV-A as the band width between 400-320 nm, UV-B between 320-280 nm, and UV-C as < 280 nm.[1] Generally, UV-C does not reach the surface of the earth because of the absorption properties of the upper atmosphere (O_2), and this is not expected to change regardless of possible alterations in the stratospheric O_3 column. The intensity and temporal patterns of ambient UV-A radiation are also not expected to be altered by possible changes in stratospheric O_3, and plants do not appear to be sensitive to this waveband in the same negative way as they are to elevated UV-B. Because of the sensitivity of many plant species to UV-B radiation, much research has been directed to that issue in growth chambers, greenhouses and ambient field plots over the past 25 years.[2] Extensive reviews of the relevant research can be found in Caldwell,[3-9] Nachtwey and Rundel,[10] National Research Council,[11-13] Teramura,[14-17] Dudek and Oppenheimer,[18] Worrest and Caldwell,[19] Tevini and Teramura,[20] Caldwell et al,[21] Tevini[22] and Rozema et al.[23]

Two large and significant research programs were completed in the U.S. during the early and mid-1970s. Fear of possible climatic effects of emissions from high flying supersonic aircraft led to the research as reported by the Climatic Impact Assessment Program (CIAP) and summarized by Caldwell.[5] Within a short time thereafter, fear of possible effects of chlorofluorocarbons on stratospheric O_3 led to the reports from the Biological and Climatic Effects Research (BACER) Program.[24-29]

Elevated Ultraviolet (UV)-B Radiation and Agriculture, by Sagar V. Krupa, Ronald N. Kickert and Hans-Jürgen Jäger.
© 1998 Springer-Verlag and Landes Bioscience.

Table 3.1. Effects of UV-B on terrestrial vegetation[a,b]

Plant characteristic	Effect
Photosynthesis	Reduced in many C3 and C4 species (at low light intensities)
Leaf conductance	Reduced (at low light intensities)
Water use efficiency	Reduced in most species
Leaf area	Reduced in many species
Specific leaf weight	Increased in many species
Crop maturation rate	Not affected
Flowering	Inhibited or stimulated
Dry matter production and yield	Reduced in many species
Sensitivity between species	Large variability in response among species
Sensitivity between cultivars (within species)	Response differs between cultivars
Drought stress sensitivity	Plants become less sensitive to UV-B, but sensitive to lack of water
Mineral stress sensitivity	Some species become less and others become more sensitive to UV-B

[a]Reprinted with permission from: Runeckles VC, Krupa SV. Environ Pollut 1994; 83:191-213. © 1994 Elsevier Science Ltd. [b]Summary conclusions from artificial UV-B exposure studies.

Primary Effects of Elevated UV-B on Crops

Krupa and Kickert,[2,30] Tevini and Teramura,[20] Tevini[22] and Runeckles and Krupa[31] have reviewed numerous artificial exposure studies conducted to determine the effects of elevated UV-B levels on plants. Table 3.1 provides a general summary of the results from those studies that is still valid in light of the most recent investigations. In their previous analysis, Krupa and Kickert[2,30] noted that very different responses had been observed for the same crop species by different investigators at different times and locations (growing conditions). In many cases, these differences also reflect cultivar and varietal differences in sensitivity within a single species.[32-34] Another cause of the differences probably comes from the use of different UV-B radiation sources, exposure systems and action spectra for computing biologically effective UV-B flux densities.[31] Action spectra play a key role in the understanding of biological impacts of UV-B, because of: (1) the differential sensitivities of various responses across the range of wavelengths in the UV-B region and (2) the magnitudes of the differences in surface flux densities at these wavelengths resulting from the shape of the O_3 absorption spectrum.[35]

The results obtained with a given species are frequently contradictory when comparing the effects of UV-B exposures in growth chambers or greenhouses to those under field conditions.[36] Such differences may well arise not only from differences in the microclimatic radiant and heat energy budgets extant in the different exposure systems at the times of exposure, but also because of the differences induced by the environmental conditions under which the plants were grown a priori. For example, although the cuticle and the epidermis may act as barriers to UV-B,[37] greenhouse-grown plants are known to have a much thinner and less well developed cuticle than field-grown plants[38] and, thus might exhibit greater sensitivity.

However, probably the factor of greatest importance in determining the relevance of many growth chamber and greenhouse studies, is the intensity of photosynthetically active radiation (PAR) to which the plants were exposed.[31] For example, Biggs et al[32] acknowledged that the sensitivities to UV-B of the soybean (*Glycine max* (L.) Merr.) cultivars they studied were enhanced by the low (one-eighth of full sunlight) PAR (photosynthetically active radiation) levels used, because of the minimal activity of photorepair processes.

The results of field studies using selective filters to remove UV-B or filter/UV-lamp combinations are themselves far from conclusive.[31] For example, Becwar et al[39] found no significant effects on dry matter accumulation of pea (*Pisum sativum* L.), potato (*Solanum tuberosum* L.), radish (*Raphanus sativus* L.) and wheat (*Triticum aestivum* L.) plants grown for 50 days at a high elevation site (3,000 m) in Colorado with high PAR and UV-B fluxes, even when lamp-filter combinations increased the effective UV-B radiation by 52%. The only significant effect observed was an early, but slight decrease in wheat plant height growth that had disappeared by the time of final harvest. In contrast, Teramura et al[33] reported net adverse effects of enhanced UV-B on the yield of the sensitive soybean cultivar, Essex, based on a 6-year study, both for individual years and when averaged over the years. However, this effect was only observed at the higher UV-B enhancement used (computed to be equivalent to a 25% stratospheric O_3 depletion). There was no adverse effect on the cultivar, Williams, when averaged over the same 6 years; indeed, a lower UV-B enhancement (equivalent to 16% O_3 depletion) resulted in a significant average increase in yield. The authors attributed the wide range of responses observed for either cultivar over the years to a strong influence of seasonal variation in microclimate.

Investigators at the University of Lancaster have used growth chambers in which UV-B enhancement was provided at high overall light intensities approximating two-thirds of full sunlight.[40] Under these artificial conditions that approach those typical of the field conditions with respect to light intensities, the vegetative growth of pea plants and their rates of photosynthesis were found to be unaffected by increased UV-B levels. However, yields were found to be somewhat depressed, presumably because of adverse effects of UV-B at various stages in the process of sexual reproduction. The vegetative growth of a range of barley (*Hordeum vulgare* L.) cultivars was unaffected by increased UV-B flux, although the cultivar, Scout, that contained little if any flavonoids, suffered visible injury. Such observations tend to confirm the conclusions of others.[31] For example, Beyschlag et al[41] found no adverse effect of UV-B on the photosynthetic rates of competing wheat and wild oat (*Avena fatua* L.) plants, although the competitiveness of wheat was increased because of UV-B-induced inhibition of the height growth of the wild oat plants. Increased flavonoid production has long been hypothesized as a protective feature of many plant species and cultivars.[42]

Sensitivity Rankings of Crop Species

Independent of some of the concerns expressed in the previous section, there is differential sensitivity in plant species exposed to enhanced UV-B. Among plant species there is evidence for a negative response (sensitive), a positive response (stimulation), and no significant difference between treatments and the control (tolerant). While a variety of plant physiological and morphological responses could be used, we choose to focus on biomass accumulation. Regardless of which response parameter is of greatest interest, and in spite of much of the plant physiological research that has been done, there is still no basis by which the sensitivity of a species or a cultivar can be determined without engaging in direct experimental work. We have updated Table 3 in Teramura.[14] We acknowledge that identifying plant species on a simple ordinal scale such as, sensitive and tolerant to enhanced UV-B contains much ambiguity because the term sensitive does not explain how sensitive in the sense of using the first derivative (rate of change) of dose-response curves. Investigators in this area of study have not yet presented dose-response curves, thus a massive recomputation of the published data would be required.

What is immediately apparent from Table 3.2 is that very different UV-B-induced responses were found for the same crop species in different studies. This situation arises due to a variety of reasons: (1) there can be intra-species differences between cultivars; for a given crop in Table 3.2, we encourage the reader to consult the references given, where there is a need to examine responses among crop cultivars; (2) different UV-B sources, flux densities and action spectra for computing biologically effective UV-B radiation, were used between various studies; and (3) often, but not always, results of studies with a given crop were observed to be reversed when comparing artificial exposure in growth chambers and greenhouses, to exposures in open, ambient field plots. An example is that of Dumpert and Knacker[36] where kohlrabi (*Brassica oleracea* L. var. *gongylodes*) showed tolerance (no response) in the greenhouse, but increased total dry weight (stimulation) under exposure in an open field. Aside from the first two reasons mentioned previously, conflicting results might have been obtained with the same crop under different exposure environments because of differences in microclimatic radiant and heat energy and moisture budgets between the two environments. Very few investigators have measured the leaf temperatures (a result of longer wavelength radiant energy at the leaf surface and latent heat flux of evapotranspiration) between the test and control plants and between the different exposure environments used.

Only two fiber crops have apparently been examined for UV-B effects on biomass accumulation, and these can be considered as tolerant (Table 3.2). Of the C3 grain crops, barley and oat (*Avena sativa* L.) are sensitive, rice (*Oryza sativa* L.) and rye (*Secale cereale* L.) are moderately sensitive and wheat and sunflower (*Helianthus annuus* L.) are tolerant. Of the C4 grain crops, we regard sweet corn (*Zea mays* L. var. *saccharata*) as sensitive, and grain sorghum (*Sorghum vulgare* Moench) as moderately sensitive. Corn and millet (*Setaria italica* (L.) Pal) appear to be tolerant to enhanced UV-B with regard to biomass accumulation. With legume seed crops, soybean is generally sensitive to UV-B, along with pea and cowpea (*Vigna sinensis* Savi). Bean (*Phaseolus* spp. L.) is moderately sensitive, and peanut (*Arachis hypogaea* L.) exhibited very mixed results, with all three types of responses (sensitive or negative; positive or

Table 3.2. Relative sensitivity of crops to INCREASED UV-B radiation based on measures of biomass accumulation[a,b,c]

Sensitivity[d]	Plant	Response effect	Exposure environment[e]	Reference
Fiber crops				
—	Cotton	Tot dry wt	gh	Ambler et al[43]
		Cotyledon dw	gh	Ambler et al[43]
		Tot dry wt	gh	Bennett[44]
Tolerant	Cotton	Crop yield	field	Hart et al[45]
		Tot dry wt	gh	Hart et al[45]
		Tot dry wt	gc	Krizek[46]
		Tot dry wt	gh, gc	Biggs and Kossuth[24]
		Tot dry wt	gh	Bennett[47]
Tolerant	Hemp (drug & fiber)	Leaf dry wt	gh	Lydon et al[48]
C3 grain crops				
—	Barley	Tot dry wt	field	Caldwell et al[49]
		Tot dry wt	gh, gc	Biggs and Kossuth[24]
		Tot dry wt	gh, gc	Hashimoto and Tajima[50]
		Tot dry wt	gh	Dumpert and Boscher[51]
		Tot dry wt	gc	Tevini et al[52]
+	Barley	Tot dry wt	gc	Tevini et al[53]
		Tot dry wt	gh	Dumpert and Knacker[36]
Tolerant	Barley	Tot dry wt	gh, gc	Tevini et al[53,54]
		Cutic. wax	gc	Steinmüeller and Tevini[37,55]
—	Oats	Tot dry wt	gh	Thai and Garrard[56]
		Tot dry wt	gh	Van and Garrard[57]
		Tot dry wt	gh	Van et al[58]
		Tot dry wt	gc	Basiouny et al[59]
+	Oats	Tot dry wt	gh, gc	Biggs and Kossuth[24]
Tolerant	Oats	Tot dry wt	gc, solarium	Biggs and Basiouny[60]
—	Rice	Tot dry wt	solarium	Biggs and Basiouny[60]
		Tot dry wt	gh, gc, field	Biggs and Kossuth[24,26]
		Crop yield	field	Biggs et al[61]
Tolerant	Rice	Tot dry wt	gh	Thai and Garrard[56]
		Tot dry wt	gh	Van et al[58]
		Tot dry wt	gh	Ambler et al[62]
		Tot dry wt	gh, gc, field	Biggs and Kossuth[24]
		Crop yield	field	Biggs and Webb[63]
—	Rye	Tot dry wt	gh	Thai and Garrard[56]
		Tot dry wt	gh	Van et al[58]
		Tot dry wt	gh, gc	Biggs and Kossuth[24]
Tolerant	Rye	Tot dry wt	gc	Biggs and Basiouny[60]
—	Wheat	Tot dry wt	gc	Hart et al[45]
		Tot dry wt	gh, gc	Biggs and Kossuth[24]
		Tot dry wt	gh, gc	Teramura[64]
		Tot dry wt	field	Webb[65]

Sensitivity[d]	Plant	Response effect	Exposure environment[e]	Reference
		Crop yield	field	Webb[65]
+	Wheat	Tot dry wt	gh, gc	Biggs and Kossuth[24,27]
		Tot dry wt	gh, gc	Dumpert and Knacker[36]
Tolerant	Wheat	Tot dry wt	gc	Krizek[46]
		Tot dry wt	gh	Ambler et al[62]
		Grain wt	gh	Ambler et al[62]
		Tot dry wt	gh	Bennett[44]
		Tot dry wt	gh, gc	Biggs and Kossuth[24]
		Tot dry wt	field	Moore et al[66]
		Crop yield	field	Moore et al[66]
		Tot dry wt	gh	Teramura[64]
		Tot dry wt	gh	Bennett[47]
		Tot dry wt	field	Becwar et al[39]
		Crop yield	field	Biggs et al[61]
		Shoot biomass	field	Gold and Caldwell[67]
		Crop yield	field	Biggs and Webb[63]
		Shoot biomass	gh, field	Barnes et al[68]
+	Sunflower	Tot dry wt	gh, gc	Biggs and Kossuth[24]
C4 grain crops				
—	Sweet corn	Tot dry wt	gh	Allen et al[69]
		Crop yield	field	Ambler et al[70]
		Plant biomass	field	Halsey et al[71]
		Tot dry wt	gh	Vu et al[72]
+	Sweet corn	Ear size	field	Halsey et al[71]
—	Sorghum	Tot dry wt	gc	Biggs and Basiouny[60]
		Tot dry wt	gh	Thai and Garrard[56]
		Tot dry wt	gh	Van et al[58]
		Tot dry wt	field	Ambler et al[70]
		Tot dry wt	gh, gc	Biggs and Kossuth[24]
Tolerant	Sorghum	Tot dry wt	field	Hart et al[45]
		Crop yield	field	Hart et al[45]
		Tot dry wt	gc	Basiouny et al[59]
—	Corn	Tot dry wt	gh, gc, field	Biggs and Kossuth[24,26]
		Crop yield	field	Biggs and Kossuth[26]
+	Corn	Tot dry wt	field	Caldwell et al[49]
		Crop yield	field	Bartholic et al[73]
		Coleoptile dw	gc	Hashimoto and Tajima[50]
		Tot dry wt	gc	Tevini et al[53]
Tolerant	Corn	Tot dry wt	gc	Biggs and Basiouny[60]
		Tot dry wt	field	Hart et al[45]
		Crop yield	field	Hart et al[45]
		Tot dry wt	gh	Thai and Garrard[56]
		Tot dry wt	gh	Van and Garrard[57]
		Tot dry wt	gh	Van et al[58]
		Tot dry wt	gc	Basiouny et al[59]
		Tot dry wt	gh, gc	Tevini et al[53,54,74]
		Crop yield	field	Biggs and Webb[63]

(cont'd.)

Sensitivity[d]	Plant	Response effect	Exposure environment[e]	Reference
—	Millet	Tot dry wt	gc	Hart et al[45]
		Tot dry wt	gh, gc	Biggs and Kossuth[24]
Tolerant	Millet	Tot dry wt	gc	Krizek[46]
		Tot dry wt	field	Hart et al[45]
		Crop yield	field	Hart et al[45]
		Tot dry wt	gh	Thai and Garrard[56]
		Tot dry wt	gh	Van and Garrard[57]
		Tot dry wt	gh	Van et al[58]
Legume seed crops				
—	Soybean	Tot dry wt	solarium	Biggs and Basiouny[60]
		Root dry wt	field	Caldwell et al[49]
		Tot dry wt	gh	Thai and Garrard[56]
		Tot dry wt	gh	Van and Garrard[57]
		Tot dry wt	gh	Van et al[58]
		Tot dry wt	gh	Allen et al[69]
		Crop yield	field	Ambler et al[70]
		Tot dry wt	gc	Basiouny et al[59]
		Tot dry wt	gh	Bennett[44]
		Tot dry wt	gh, gc	Biggs and Kossuth[24,25,27]
		Biomass	gc	Kossuth and Biggs[75]
		Tot dry wt	gh	Vu et al[72]
		Tot dry wt	gh	Teramura[64]
		Tot dry wt	gh, gc	Biggs et al[32]
		Tot dry wt	gh	Vu et al[76]
		Tot dry wt	gh, gc	Teramura and Perry[77]
		Tot dry wt	gh, gc	National Research Council[13]
		Tot dry wt	gh	Teramura et al[78]
		Tot dry wt	gh	Murali and Teramura[79]
—	Soybean	Tot dry wt	field	Lydon et al[80]
		Tot dry wt	field	Murali and Teramura[81]
		Crop yield	field	Teramura[17]
		Tot dry wt	gh, field	Teramura and Murali[82]
		Tot dry wt	gh	Murali and Teramura[83]
		Tot dry wt	gh	Teramura and Sullivan[84]
		Tot dry wt	gh	Murali et al[85]
		Crop yield	field	Teramura and Sullivan[86]
+	Soybean	Crop yield	field	Teramura and Sullivan[86]
Tolerant	Soybean	Tot dry wt	gc	Biggs and Basiouny[60]
		Tot dry wt	field	Hart et al[45]
		Crop yield	field	Hart et al[45]
		Tot dry wt	gc	Krizek[46]
		Tot dry wt	gh	Bennett[47]
		Crop yield	field	Biggs et al[61]
		Tot biomass	gh	Teramura[87]
		Crop yield	field	Biggs and Webb[63]
		Crop yield	field	Murali and Teramura[88]
		Tot dry wt	field	Murali and Teramura[81]

Sensitivity[d]	Plant	Response effect	Exposure environment[e]	Reference
		Tot dry wt	gh, field	Teramura and Murali[82]
—	Pea	Tot dry wt	gc, solarium	Biggs and Basiouny[60]
		Tot dry wt	gc	Hart et al[45]
		Tot dry wt	gh	Thai and Garrard[56]
		Tot dry wt	gc	Krizek et al[89]
		Tot dry wt	gh	Van and Garrard[57]
		Tot dry wt	gh	Van et al[58]
		Tot dry wt	gh, gc	Brandle et al[90]
		Tot dry wt	gh	Allen et al[69]
		Tot dry wt	gh, gc, field	Biggs and Kossuth[24,26]
		Crop yield	field	Biggs and Kossuth[26]
		Tot dry wt	gh	Vu et al[72]
		Tot dry wt	gc	Basiouny[91]
		Tot dry wt	gh, gc	Vu et al[92]
+	Pea	Tot dry wt	gh, gc	Biggs and Kossuth[24]
		Biomass	gc	Kossuth and Biggs[75]
Tolerant	Pea	Tot dry wt	gc, gh	Biggs and Basiouny[60]
		Tot dry wt	field	Fox and Caldwell[93]
		Tot dry wt	field	Moore et al[66]
		Tot dry wt	gc	Basiouny[91]
		Tot dry wt	field	Becwar et al[39]
—	Cowpeas	Tot dry wt	gc	Biggs and Basiouny[60]
		Tot dry wt	field	Biggs and Kossuth[24]
		Crop yield	field	Biggs and Kossuth[26]
		Biomass	gc	Kossuth and Biggs[75]
Tolerant	Cowpeas	Tot dry wt	gh	Biggs and Basiouny[60]
—	Beans	Tot dry wt	gc	Biggs and Basiouny[60]
		Tot dry wt	gh	Bennett[44]
		Tot dry wt	gh, gc	Biggs and Kossuth[24]
		Biomass	gc	Kossuth and Biggs[75]
		Tot dry wt	gc	Tevini et al[53]
		Tot dry wt	gc	Basiouny[91]
		Tot dry wt	gh	Dumpert and Boscher[51]
		Prim leaf dw	gc	Tevini et al[74]
		Tot dry wt	gh, gc	Dumpert and Knacker[36]
+	Beans	Crop yield	field	Bartholic et al[73]
Tolerant	Beans	Tot dry wt	gh, gc	Biggs and Basiouny[60]
		Crop yield	gh	Hart et al[45]
		Tot dry wt	gc	Krizek[46]
		Tot dry wt	field	Ambler et al[70]
		Crop yield	field	Ambler et al[70]
		Tot dry wt	gh	Bennett[47]
		Tot dry wt	gh, gc	Tevini et al[52]
—	Peanut	Tot dry wt	gc	Hart et al[45]
		Crop yield	field	Biggs and Kossuth[26]
		Tot dry wt	field	Biggs and Kossuth[26]
+	Peanut	Tot dry wt	gh, gc	Biggs and Kossuth[24]
		Tot dry wt	field	Biggs and Kossuth[26]

(cont'd.)

Sensitivity[d]	Plant	Response effect	Exposure environment[e]	Reference
		Biomass	gc	Kossuth and Biggs[75]
Tolerant	Peanut	Tot dry wt	gc, solarium	Biggs and Basiouny[60]
		Tot dry wt	field	Hart et al[45]
		Crop yield	field	Hart et al[45]
		Tot dry wt	gh	Thai and Garrard[56]
		Tot dry wt	gh	Van and Garrard[57]
		Tot dry wt	gh	Van et al[58]
		Tot dry wt	gc	Basiouny et al[59]
Fruit crops				
—	Tomato	Tot dry wt	gh	Biggs and Basiouny[60]
		Tot dry wt	gc	Hart et al[45]
		Tot dry wt	gh	Thai and Garrard[56]
		Tot dry wt	gh	Van et al[58]
		Tot dry wt	gh, gc, field	Biggs and Kossuth[24,26]
		Crop yield	field	Biggs and Kossuth[26]
		Plant biomass	field	Halsey et al[71]
		Crop yield	field	Halsey et al[71]
		Crop yield	field	Nachtwey and Rundel[10]
+	Tomato	Crop yield	gh, gc	Prudot and Basiouny[94]
Tolerant	Tomato	Crop yield	field	Bartholic et al[73]
		Tot dry wt	gc	Biggs and Basiouny[60]
		Tot dry wt	field	Caldwell et al[49]
		Crop yield	field	Hart et al[45]
		Tot dry wt	gc	Krizek[46]
		Tot dry wt	gc	Basiouny[91]
—	Cucumber	Tot dry wt	gc	Biggs and Basiouny[60]
		Crop yield	gc	Nakazawa et al[95]
		Leaf dry wt	gh	Ambler et al[62]
		Tot dry wt	gh	Bennett[44]
		Tot dry wt	gh, gc	Biggs and Kossuth[24]
		Tot dry wt	gh	Krizek[96,97]
		Cotyledon dw	gc	Hashimoto and Tajima[50]
		Tot dry wt	gh	Bennett[47]
		Tot dry wt	gc	Basiouny[91]
		Cotyledon dw	gc	Tevini et al[74]
				National Research Council[13]
				Steinmüeller and Tevini[37]
		Cutic. wax	gc	Murali and Teramura[98]
		Tot dry wt	gh	Steinmüeller and Tevini[55]
		Cutic. wax	gc	Tevini and Iwanzik[99]
		Tot dry wt	gc	
Tolerant	Cucumber	Tot dry wt	gh	Biggs and Basiouny[60]
		Tot dry wt	gc	Krizek[46]
		Tot dry wt	gh	Murali and Teramura[98]
—	Squash	Tot dry wt	gc	Biggs and Basiouny[60]
		Crop yield	field	Ambler et al[70]
		Tot dry wt	field	Ambler et al[70]
		Tot dry wt	gh, gc	Biggs and Kossuth[24]

Sensitivity[d]	Plant	Response effect	Exposure environment[e]	Reference
		Crop yield	field	Biggs and Kossuth[26]
		Tot dry wt	gc	Basiouny[91]
—	Okra	Tot dry wt	gh, gc	Biggs and Kossuth[24]
—	Pumpkin	Tot dry wt	gh, gc	Biggs and Kossuth[24]
—	Watermelon	Tot dry wt	gh, gc	Biggs and Kossuth[24,25]
—	Cantaloupe	Crop quality	field	Lipton[100]
		Tot dry wt	gh, gc	Biggs and Kossuth[24]
		Crop quality	field	Lipton and O'Grady[101]
—	Red raspberry	Crop yield	gh	Renquist et al[102]
—	Blueberry	Crop yield	gh	Biggs and Kossuth[28]
		Crop yield	gh	Kossuth and Biggs[103]
		Crop yield	gh	Kossuth and Biggs[104]
—	Pepper	Tot dry wt	field	Caldwell et al[49]
		Crop yield	field	Hart et al[45]
Tolerant	Pepper	Crop yield	field	Hart et al[45]
		Tot dry wt	gh, gc	Biggs and Kossuth[24]
—	Eggplant	Cotyledon dw	gc	Hashimoto and Tajima[50]
+	Eggplant	Tot dry wt	gh, gc	Biggs and Kossuth[24]
Tolerant	Orange	Biomass growth	field	Biggs and Kossuth[29]

Vegetable flower crops

Sensitivity[d]	Plant	Response effect	Exposure environment[e]	Reference
—	Cauliflower	Tot dry wt	gh, gc	Biggs and Kossuth[24]
—	Broccoli	Tot dry wt	gh, gc	Biggs and Kossuth[24]
		Tot dry wt	field	Ambler et al[70]
		Crop yield	field	Ambler et al[70]
+	Artichoke	Tot dry wt	gh, gc	Biggs and Kossuth[24]

Ornamental flower crops

Sensitivity[d]	Plant	Response effect	Exposure environment[e]	Reference
—	Bluebell	Tot dry wt	gh	Krizek and Semeniuk[105]
—	Ivy geranium	Leaf area	gh	Rangarajan and Tibbitts[106]
Tolerant	Richardson geranium	Shoot dry wt	field	Caldwell et al[49]
—	Marigold	Tot dry wt	gc	Hart et al[45]
Tolerant	Marigold	Flower number	field	Hart et al[45]
+	Yellow alyssum	Tot dry wt	field	Fox and Caldwell[93]
		Shoot biomass	field	Gold and Caldwell[67]
Tolerant	Yellow alyssum	Tot dry wt	field	Fox and Caldwell[93]
+	Floribunda rose	Petal color	in vitro	Maekawa et al[107]
Tolerant	Poinsettia	Tot dry wt	gh	Semeniuk and Stewart[108]
—	Coleus	Leaf discolor	field	Hart et al[45]
		Tot dry wt	gh	Hart et al[45]
Tolerant	Coleus	Tot dry wt	gh	Semeniuk and Stewart[109]

(cont'd.)

Sensitivity[d]	Plant	Response effect	Exposure environment[e]	Reference
—	Petunia	Tot dry wt	gc	Hart et al[45]
Tolerant	Petunia	Flower number	field	Hart et al[45]
Tolerant	Chrysanthemum	Flower number	field	Hart et al[45]
Leaf crops				
—	Collards	Tot dry wt	gh	Thai and Garrard[56]
		Tot dry wt	gh	Van et al[58]
		Tot dry wt	gc	Basiouny et al[59]
		Tot dry wt	gh, gc	Biggs and Kossuth[24]
		Tot dry wt	gc	Basiouny[91]
Tolerant	Collards	Tot dry wt	gc, gh	Biggs and Basiouny[60]
—	Chards	Tot dry wt	gh, gc	Biggs and Kossuth[24]
—	Brussels sprouts	Tot dry wt	gh, gc	Biggs and Kossuth[24]
—	Kale	Tot dry wt	gh, gc	Biggs and Kossuth[24]
—	Mustard	Tot dry wt	gh, gc, field	Biggs and Kossuth[24]
		Crop yield	field	Biggs and Kossuth[26]
		Shoot biomass	gh	Gold and Caldwell[67]
—	White mustard	Tot dry wt	field	Bogenrieder and Klein[110]
—	Spinach	Tot dry wt	gh	Dumpert and Boscher[51]
		Tot dry wt	gh	Dumpert and Knacker[36]
		Shoot biomass	gh	Gold and Caldwell[67]
—	Lettuce	Tot dry wt	gh, gc	Hart et al[45]
		Tot dry wt	gh, gc	Biggs and Kossuth[24]
		Tot dry wt	gh, gc	Bogenrieder and Douté[111]
		Tot dry wt	gh, gc	Dumpert and Knacker[36]
Tolerant	Lettuce	Tot dry wt	solarium	Biggs and Basiouny[60]
		Tot dry wt	gc	Krizek[46]
—	Cabbage	Tot dry wt	gh	Thai and Garrard[56]
		Tot dry wt	gh	Van et al[58]
		Tot dry wt	gh, gc	Biggs and Kossuth[24]
+	Cabbage	Tot fresh wt	field	Dumpert and Knacker[36]
Tolerant	Cabbage	Tot dry wt	gh	Biggs and Basiouny[60]
		Tot dry wt	field	Hart et al[45]
		Crop yield	field	Hart et al[45]
		Tot dry wt	gh	Dumpert and Knacker[36]
—	Kohlrabi	Tot dry wt	gh, gc	Biggs and Kossuth[24]
+	Kohlrabi	Tot dry wt	field	Dumpert and Knacker[36]
Tolerant	Kohlrabi	Tot dry wt	gh	Dumpert and Knacker[36]
—	Alyce clover	Biomass	gc	Kossuth and Biggs[75]
—	Clover	Biomass	gh	Biggs and Kossuth[24]
Tolerant	Alpine (whiproot) clover	Shoot yield	field	Caldwell[3]
Tolerant	Clover	Tot dry wt	gh	Bennett[44,47]
Tolerant	Red clover	Tot dry wt	field	Fox and Caldwell[93]
—	Alfalfa	Tot dry wt	gc, gh	Hart et al[45]
		Shoot biomass	field	Fox and Caldwell[93]

Sensitivity[d]	Plant	Response effect	Exposure environment[e]	Reference
+	Alfalfa	Shoot biomass	field	Gold and Caldwell[67]
Tolerant	Alfalfa	Tot dry wt	field	Caldwell et al[49]
		Tot dry wt	gh	Ambler et al[62]
		Tot dry wt	field	Fox and Caldwell[93]
Tolerant	Kentucky bluegrass	Tot dry wt	field	Fox and Caldwell[93]
Tolerant	Bermuda grass	Crop yield	field	Hart et al[45]
Tolerant	Orchard grass	Crop yield	field	Hart et al[45]
		Tot dry wt	gh	Hart et al[45]
Tolerant	Digit grass	Tot dry wt	gh	Thai and Garrard[56]
		Tot dry wt	gh	Van and Garrard[57]
		Tot dry wt	gh	Van et al[58]
Tolerant	Tobacco	Tot dry wt	gc, solarium	Biggs and Basiouny[60]
		Tot dry wt	field	Hart et al[45]
		Crop yield	field	Hart et al[45]

Stem crops

Sensitivity[d]	Plant	Response effect	Exposure environment[e]	Reference
—	Rhubarb	Tot dry wt	gh, gc	Biggs and Kossuth[24]
—	Sugarcane	Tot dry wt	gh	Elawad et al[112]
		Crop yield	gh	
+	Celery	Tot dry wt	gh, gc	Biggs and Kossuth[24]
Tolerant	Celery	Tot dry wt	solarium	Biggs and Basiouny[60]
Tolerant	Asparagus	Tot dry wt	gh, gc	Biggs and Kossuth[24]

Root, bulb and tuber crops

Sensitivity[d]	Plant	Response effect	Exposure environment[e]	Reference
—	Sugar beet	Tot dry wt	gc	Hart et al[45]
		Tot dry wt	field	Ambler et al[70]
		Shoot biomass	gh	Gold and Caldwell[67]
—	Carrot	Tot dry wt	gc	Biggs and Basiouny[60]
		Tot dry wt	gc	Hart et al[45]
		Tot dry wt	gc	Basiouny[91]
+	Carrot	Tot dry wt	gh, gc	Biggs and Kossuth[24]
Tolerant	Carrot	Tot dry wt	gh	Biggs and Basiouny[60]
—	Rutabaga	Tot dry wt	gh, gc	Biggs and Kossuth[24]
—	Turnip	Tot dry wt	solarium	Biggs and Basiouny[60]
		Tot dry wt	field	Inagaki et al[113]
		Crop yield		
—	Potato	Tot dry wt	field	Halsey et al[71]
		Crop yield	field	Halsey et al[71]
+	Potato	Tot dry wt	field	Biggs and Kossuth[26]
		Tot dry wt	field	Halsey et al[71]
Tolerant	Potato	Crop yield	field	Biggs and Kossuth[26]
		Tot dry wt	field	Moore et al[66]
		Crop yield	field	Moore et al[66]
		Tot dry wt	field	Becwar et al[39]
—	Radish	Tot dry wt	gc, gh	Biggs and Basiouny[60]

(cont'd.)

Sensitivity[d]	Plant	Response effect	Exposure environment[e]	Reference
		Tot dry wt	gh, gc	Hart et al[45]
		Cotyledon dw	gc	Hashimoto and Tajima[50]
		Tot dry wt	gc	Basiouny[91]
		Tot dry wt	gh, gc	Tevini et al[52,114]
		Shoot biomass	gh	Gold and Caldwell[67]
		Cotyledon fresh wt	gc	Iwanzik[115]
+	Radish	Tot dry wt	gh, gc, field	Biggs and Kossuth[24,26]
		Tot dry wt	gc	Tevini et al[53]
Tolerant	Radish	Tot dry wt	solarium	Biggs and Basiouny[60]
		Tot dry wt	gc	Krizek[46]
		Crop yield	field	Biggs and Kossuth[26]
		Tot dry wt	field	Moore et al[66]
		Crop yield	field	Moore et al[66]
		Tot dry wt	gh, gc	Tevini et al[53]
		Tot dry wt	field	Becwar et al[39]
		Cotyledon dw	gc	Tevini et al[74]
		Tot dry wt	gh, gc	Dumpert and Knacker[36]
+	Chufa	Tot dry wt	gh, gc	Biggs and Kossuth[24]
—	Onion	Tot dry wt	gc	Biggs and Basiouny[60]
		Tot dry wt	field	Fox and Caldwell[93]
Tolerant	Onion	Tot dry wt	gh	Biggs and Basiouny[60]
		Crop yield	gh	Hart et al[45]
		Tot dry wt	gh, gc	Biggs and Kossuth[24]
		Tot dry wt	gc	Basiouny[91]
Tolerant	Parsnip	Tot dry wt	gh, gc	Biggs and Kossuth[24]

[a]Reprinted with permission from: Krupa SV, Kickert RN. Environ Pollut 1989; 61(4):263-393. © 1989 Elsevier Science Ltd. [b]Reference to total dry weight does not necessarily refer to end-of-season, and in many cases, is often after only a few days, or weeks, of growth. [c]After Table 3 in Teramura[14] and updated to 1988. [d]Response showing a decrease under UV-B is "–", showing an increase is "+", and showing relatively little change is "tolerant." [e]gh, greenhouse; gc, growth chamber.

stimulation; or no response or tolerance) having been observed in field exposures. The fruit crops probably exhibit the largest variety of sensitive species: tomato (*Lycopersicon esculentum* Miller), cucumber (*Cucumis sativus* L.), squash (*Cucurbita* spp. L.), okra (*Hibiscus esculentus* L.), pumpkin (*Cucurbita pepo* L.), melon (*Cucumis melo* L.), red raspberry (*Rubus strigosus* L.) and blueberry (*Vaccinium* spp. L.). Pepper (*Capsicum frutescens* L.) showed mixed results, and we consider eggplant (*Solanum melongena* L.) and orange (*Citrus* spp. L.) to be tolerant. Of the few vegetable flower crops for which little if any replication of original research has been performed, both cauliflower (*Brassica oleracea* L. var. *botrytis*) and broccoli (*Brassica oleracea* L. var. *botrytis*) are sensitive to increased UV-B, while artichoke (*Cynara scolymus* L.) appears to be tolerant.

It can be misleading to list ornamental flower crops in Table 3.2, where the plant response used as a frame of reference is biomass accumulation. Many of these plants have a market value based upon visual appearance rather than size or weight of the plant. In general, the ornamental plants listed in the table appear to display tolerance to increased UV-B.

Of the leaf crops, we consider collards (*Brassica oleracea* L. var. *acephala*), chards (*Beta vulgaris* L. var. *cicla* (L.) Koch), brussels sprouts (*Brassica oleracea* L. var. *gemmifera*), kale (*Brassica oleracea* L. var. *acephala*), the mustards (*Brassica* spp. L.) and spinach (*Spinacia oleracea* L.) to be sensitive, with lettuce (*Lactuca sativa* L.) being moderately sensitive. With emphasis on field results, we consider cabbage (*Brassica oleracea* L. var. *capitata*), kohlrabi, most of the clovers (*Trifolium* spp. L.), and alfalfa (*Medicago sativa* L.) to be tolerant. Several grasses also appear to be tolerant such as Kentucky bluegrass (*Poa pratensis* L.), Bermuda grass (*Cynodon dactylon* (L.) Pers.), orchard grass (*Dactylis glomerata* L.) and digit grass (*Digitaria decumbens* Stent). The only evidence available shows tobacco (*Nicotiana tabacum* L.) to be tolerant to enhanced UV-B. Among the stem crops, rhubarb (*Rheum rhaponticum* L.) and sugarcane (*Saccharum officinarum* L.) might be sensitive, but there is no field evidence. The only evidence appears to show that celery (*Apium graveolens* L.) and asparagus (*Asparagus officinalis* L.) do not respond negatively to enhanced UV-B.

Of the root, bulb and tuber crops, sugar beet (*Beta vulgaris* L.), carrot (*Daucus carota* L.), rutabaga (*Brassica napus* L. var. *napobrassica*) and turnip (*Brassica rapa* L.) are considered to be sensitive. Of all the evidence examined, potato is the only crop for which multiple tests were performed with ambient field exposures. Based on the results obtained, we consider this crop as a whole to vary from moderately sensitive to tolerant, depending upon the cultivar and weather conditions. Radish, onion (*Allium cepa* L.) and parsnip (*Pastinaca sativa* L.) are considered to be tolerant, although convincing field evidence is lacking for the last two crops. Chufa (*Cyperus esculentus* L.), the tuberous roots of a sedge consumed by people in southern Europe, did not show a negative response in the one and only artificial exposure on record.

Table 3.3 presents a classification by the relative sensitivity of cultivated crops exposed to enhanced UV-B radiation with regard to biomass accumulation.

Table 3.3. Summary of relative sensitivity of crops to increased UV-B radiation based on measures of biomass accumulation[a]

Crop category	Sensitive	Moderately sensitive	Moderately sensitive to tolerant	Tolerant
Fiber	—	—	—	Cotton
	—	—	—	Hemp (drug & fiber)
C3 grain	Barley	Rice	—	Wheat
	Oats	Rye	—	Sunflower
C4 grain	Sweet corn	Sorghum	—	Corn
	—	—	—	Millet
Legume seed	Soybean	Beans	Peanut	—
	Pea	—	—	—
	Cowpeas	—	—	—
Fruit	Tomato	Pepper	—	Eggplant
	Cucumber	—	—	Orange
	Squash	—	—	—
	Okra	—	—	—
	Pumpkin	—	—	—
	Watermelon	—	—	—
	Cantaloupe	—	—	—
	Red raspberry	—	—	—
	Blueberry	—	—	—
Vegetable, flower	Cauliflower	—	—	Artichoke
	Broccoli	—	—	—
Ornamental, flower	Bluebell	Petunia	—	Richardson geranium
	Coleus	—	—	Marigold
	Ivy geranium	—	—	Yellow alyssum
	—	—	—	Floribunda rose
	—	—	—	Poinsettia
	—	—	—	Chrysanthemum

Crop category	Sensitive	Moderately sensitive	Moderately sensitive to tolerant	Tolerant
Leaf	Collards	Lettuce	—	Cabbage
	Chards	—	—	Kohlrabi
	Brussel sprouts	—	—	Alpine (whiproot) clover
	Kale	—	—	Clover
	Mustard	—	—	Red clover
	White mustard	—	—	Alfalfa
	Spinach	—	—	Kentucky bluegrass
	Alyce clover	—	—	Bermuda grass
	Clover	—	—	Orchard grass
	—	—	—	Digit grass
	—	—	—	Tobacco
Stem	Rhubarb	—	—	Celery
	Sugarcane	—	—	Asparagus
Root, bulb & tuber	Sugar beet	—	Potato	Radish
	Carrot	—	—	Chufa
	Rutabaga	—	—	Onion
	Turnip	—	—	Parsnip

[a]Reprinted with permission from: Krupa SV, Kickert RN. Environ Pollut 1989; 61(4):263-393. © 1989 Elsevier Science Ltd.

References

1. German Bundestag. Ozone depletion, changes in UV-B radiation and their effects. In: Protecting the Earth—A Status Report with Recommendations for a New Energy Policy. Bonn, Germany: German Bundestag, 1992:571-616.
2. Krupa SV, Kickert RN. The Effects of Elevated Ultraviolet (UV)-B Radiation on Agricultural Production. A peer reviewed critical assessment report submitted to the Formal Commission on "Protecting the Earth's Atmosphere" of the German Parliament, Bonn, Federal Republic of Germany, 1993.
3. Caldwell MM. Solar ultraviolet radiation as an ecological factor for alpine plants. Ecol Monogr 1968; 38(3):243-268.
4. Caldwell MM. Solar UV irradiation and the growth and development of higher plants. In: Giese AC, ed. Photophysiology, Vol. VI. Current Topics in Photobiology and Photochemistry. New York: Academic Press, 1971:131-177.
5. Caldwell MM. Summary—CIAP Monograph 5—Impacts of climatic change on the biosphere. In: Grobecker AJ, Coroniti SC, Cannon Jr RH, eds. Report of Findings—The Effects of Stratospheric Pollution by Aircraft. DOT-TST-75-50. Washington, DC: Climatic Impact Assessment Program, U.S. Dept. of Transportation, 1974:G-1-G-80 (Appendix G).

6. Caldwell MM. The effects of solar UV-B radiation (280-315 nm) on higher plants: Implications of stratospheric ozone reduction. In: Castellani A, ed. Research in Photobiology. London: Plenum, 1977:597-607.
7. Caldwell MM. Plant life and ultraviolet radiation: Some perspective in the history of the earth's UV climate. BioScience 1979; 29(9):520-525.
8. Caldwell MM. Plant responses to solar ultraviolet radiation. In: Lange OL, Nobel PS, Osmond CB et al, eds. Encyclopedia of Plant Physiology, New Series, Vol. 12A, Physiological Plant Ecology. Berlin: Springer-Verlag, 1981:169-197.
9. Caldwell MM. Solar UV radiation as a selective force in the evolution of terrestrial plant life. In: Calkins J, ed. The Role of Solar Ultraviolet Radiation in Marine Ecosystems. New York: Plenum Pub Corp, 1982:663-675.
10. Nachtwey DS, Rundel RD. Ozone change: Biological effects. In: Bower FA, Ward RB, eds. Stratospheric Ozone and Man. Boca Raton: CRC Press, Inc., 1982:81-121.
11. National Research Council. Ecosystems and their components. In: Causes and Effects of Stratospheric Ozone Reduction: An Update. Washington, DC: National Academy Press, 1982:62-74.
12. National Research Council. Effects of UV-B radiation on plants and vegetation as ecosystem components. In: Causes and Effects of Changes in Stratospheric Ozone: Update 1983. Washington, DC: National Academy Press, 1984:206-217.
13. National Research Council. Effects on biota—introduction and summary. In: Causes and Effects of Changes in Stratospheric Ozone: Update 1983. Washington, DC: National Academy Press, 1984:135-143.
14. Teramura AH. Effects of ultraviolet-B radiation on the growth and yield of crop plants. Physiol Plant 1983; 58:415-427.
15. Teramura AH. Interaction between UV-B radiation and other stresses in plants. In: Worrest RC, Caldwell MM, eds. Stratospheric Ozone Reduction, Solar Ultraviolet Radiation and Plant Life; Workshop on the Impact of Solar Ultraviolet Radiation Upon Terrestrial Ecosystems: 1. Agricultural Crops, 27-30 September 1983, Bad Windsheim, West Germany. New York: Springer-Verlag, 1986:327-343.
16. Teramura AH. Overview of our current state of knowledge of UV effects on plants. In: Titus JG, ed. Effects of Changes in Stratospheric Ozone and Global Climate, Vol. 1: Overview. Washington, DC: U.S. EPA and United Nations Environment Programme, 1986:165-173.
17. Teramura AH. The potential consequences of ozone depletion upon global agriculture. In: Titus JG, ed. Effects of Changes in Stratospheric Ozone and Global Climate, Vol. 2: Stratospheric Ozone. Washington, DC: U.S. EPA and United Nations Environment Programme, 1986:255-262.
18. Dudek DJ, Oppenheimer M. The implications of health and environmental effects for policy. In: Titus JG, ed. Effects of Changes in

Stratospheric Ozone and Global Climate, Vol. 1: Overview. Washington, DC: U.S. EPA and United Nations Environment Programme, 1986:357-379.
19. Worrest RC, Caldwell MM, eds. Stratospheric Ozone Reduction, Solar Ultraviolet Radiation and Plant Life; NATO Advanced Research Workshop on the Impact of Solar Ultraviolet Radiation Upon Terrestrial Ecosystems: 1. Agricultural Crops; 27-30 September 1983, Bad Windsheim, West Germany. NATO ASI Series, Series G, Ecological Sciences, Vol. 8. Berlin: Springer-Verlag, 1986.
20. Tevini M, Teramura AH. UV-B effects on terrestrial plants. Photochem Photobiol 1989; 50(4):479-487.
21. Caldwell MM, Teramura AH, Tevini M. The changing solar ultraviolet climate and the ecological consequences for higher plants. Trends Ecol Evolut 1989; 4:363-367.
22. Tevini M, ed. UV-B Radiation and Ozone Depletion: Effects on Humans, Animals, Plants, Microorganisms and Materials. Boca Raton: Lewis Publishers, 1993.
23. Rozema J, Gieskes WWC, van der Geijn SC et al, eds. Special Issue: UV-B and Biosphere. Plant Ecol 1997; 128:1-313.
24. Biggs RH, Kossuth SV. Impact of solar UV-B radiation on crop productivity—Effects of ultraviolet-B radiation enhancements on eighty-two different agricultural species. In: UV-B Biological and Climatic Effects Research (BACER), FY 77-78 Research Report on the Impacts of Ultraviolet-B Radiation on Biological Systems: A Study Related to Stratospheric Ozone Depletion. Final Report, Vol. II, SIRA File No.142.23, EPA-IAG-D6-0168. Washington, DC: USDA EPA, Stratospheric Impact Research and Assessment Program (SIRA), U.S. EPA, 1978.
25. Biggs RH, Kossuth SV. Impact of solar UV-B radiation on crop productivity—Effects of ultraviolet-B radiation enhancements on soybean and watermelon varieties. In: UV-B Biological and Climatic Effects Research (BACER), FY 77-78 Research Report on the Impacts of Ultraviolet-B Radiation on Biological Systems: A Study Related to Stratospheric Ozone Depletion. Final Report, Vol. II, SIRA File No.142.23, EPA-IAG-D6-0168. Washington, DC: USDA EPA, Stratospheric Impact Research and Assessment Program (SIRA), U.S. EPA, 1978.
26. Biggs RH, Kossuth SV. Impact of solar UV-B radiation on crop productivity—Effects of ultraviolet-B radiation enhancements under field conditions on potatoes, tomatoes, corn, rice, southern peas, peanuts, squash, mustard and radish. In: UV-B Biological and Climatic Effects Research (BACER), FY 77-78 Research Report on the Impacts of Ultraviolet-B Radiation on Biological Systems: A Study Related to Stratospheric Ozone Depletion. Final Report, Vol. II, SIRA File No.142.23, EPA-IAG-D6-0168. Washington, DC: USDA EPA, Stratospheric Impact Research and Assessment Program (SIRA), U.S. EPA, 1978.

27. Biggs RH, Kossuth SV. Impact of solar UV-B radiation on crop productivity—Effects of ultraviolet-B radiation enhancements and PAR flux densities on several growth parameters as related to NCE, dark respiration, and transpiration of soybean and several growth parameters of wheat. In: UV-B Biological and Climatic Effects Research (BACER), FY 77-78 Research Report on the Impacts of Ultraviolet-B Radiation on Biological Systems: A Study Related to Stratospheric Ozone Depletion. Final Report, Vol. II, SIRA File No.142.23, EPA-IAG-D6-0168. Washington, DC: USDA EPA, Stratospheric Impact Research and Assessment Program (SIRA), U.S. EPA, 1978.
28. Biggs RH, Kossuth SV. Impact of solar UV-B radiation on crop productivity—Effects of ultraviolet-B radiation enhancement on reproduction and vegetative growth of blueberry. In: UV-B Biological and Climatic Effects Research (BACER), FY 77-78 Research Report on the Impacts of Ultraviolet-B Radiation on Biological Systems: A Study Related to Stratospheric Ozone Depletion. Final Report, Vol. II, SIRA File No.142.23, EPA-IAG-D6-0168. Washington, DC: USDA EPA, Stratospheric Impact Research and Assessment Program (SIRA), U.S. EPA, 1978.
29. Biggs RH, Kossuth SV. Impact of solar UV-B radiation on crop productivity—Effects of ultraviolet-B radiation enhancement on reproduction and vegetative growth of citrus. In: UV-B Biological and Climatic Effects Research (BACER), FY 77-78 Research Report on the Impacts of Ultraviolet-B Radiation on Biological Systems: A Study Related to Stratospheric Ozone Depletion. Final Report, Vol. II, SIRA File No.142.23, EPA-IAG-D6-0168. Washington, DC: USDA EPA, Stratospheric Impact Research and Assessment Program (SIRA), U.S. EPA, 1978.
30. Krupa SV, Kickert RN. The Greenhouse Effect: Impacts of ultraviolet-B (UV-B) radiation, carbon dioxide (CO_2), and ozone (O_3) on vegetation. Environ Pollut 1989; 61(4):263-393.
31. Runeckles VC, Krupa SV. The impact of UV-B and ozone on terrestrial vegetation. Environ Pollut 1994; 83:191-213.
32. Biggs RH, Kossuth SV, Teramura AH. Response of 19 cultivars of soybeans to ultraviolet-B irradiance. Physiol Plant 1981; 53(1):19-26.
33. Teramura AH, Sullivan JH, Lyden J. Effects of UV-B radiation on soybean yield and seed quality: A 6-year field study. Physiol Plant 1990; 80:5-11.
34. Teramura AH, Ziska LH, Sztein AE. Changes in growth and photosynthetic capacity of rice with increased UV-B radiation. Physiol Plant 1991; 83:373-380.
35. Caldwell MM, Camp LB, Warner CW et al. Action spectra and their key role in assessing biological consequences of solar UV-B radiation change. In: Worrest RC, Caldwell MM, eds. Stratospheric Ozone Reduction, Solar Ultraviolet Radiation and Plant Life; Workshop on the Impact of Solar Ultraviolet Radiation Upon Terrestrial Ecosystems: 1. Agricultural Crops, 27-30 September 1983, Bad Windsheim, West Germany. New York: Springer-Verlag, 1986:87-111.

36. Dumpert K, Knacker T. A comparison of the effects of enhanced UV-B radiation on some crop plants exposed to greenhouse and field conditions. Biochem Physiol Pflanz 1985; 180(8):599-612.
37. Steinmüeller D, Tevini M. Action of ultraviolet radiation (UV-B) upon cuticular waxes in some crop plants. Planta (Berl) 1985; 164(4):557-564.
38. Martin JT, Juniper BE. The Cuticles of Plants. New York: St. Martins, 1970.
39. Becwar M, Moore III FD, Burke MJ. Effects of deletion and enhancement of ultraviolet-B (280-315 nm) radiation on plants grown at 3000 m elevation. J Amer Soc Hort Sci 1982; 107(5):771-774.
40. Hatcher PE, Paul ND. The effect of elevated UV-B radiation on herbivory of pea by *Autographa gamma*. Entomol Exp Appl 1994; 71:227-233.
41. Beyschlag W, Barnes PW, Flint SD et al. Enhanced UV-B radiation has no effect on photosynthetic characteristics of wheat (*Triticum aestivum* L.) and wild oat (*Avena fatua* L.) under greenhouse and field conditions. Photosynthetica 1988; 22:516-525.
42. Beggs CJ, Schneider-Ziebert U, Wellmann E. UV-B radiation and adaptive mechanisms in plants. In: Worrest RC, Caldwell MM, eds. Stratospheric Ozone Reduction, Solar Ultraviolet Radiation and Plant Life; Workshop on the Impact of Solar Ultraviolet Radiation Upon Terrestrial Ecosystems: 1. Agricultural Crops, 27-30 September 1983, Bad Windsheim, West Germany. New York: Springer-Verlag, 1986:235-250.
43. Ambler JE, Krizek DT, Semeniuk P. Influence of UV-B radiation on early seedling growth and translocation of ^{65}Zinc from cotyledons in cotton. Physiol Plant 1975; 34(3):177-181.
44. Bennett JH. Effects of UV-B radiation on photosynthesis and growth of selected agricultural crops. In: UV-B Biological and Climatic Effects Research (BACER), FY 77-78 Research Report on the Impacts of Ultraviolet-B Radiation on Biological Systems: A Study Related to Stratospheric Ozone Depletion. Final Report, Vol. I, SIRA File No.142.21, EPA-IAG-D6-0168. Washington, DC: USDA EPA, Stratospheric Impact Research and Assessment Program (SIRA), U.S. EPA, 1978.
45. Hart RH, Carlson GE, Klueter HH et al. Response of economically valuable species to ultraviolet radiation. In: Nachtwey DS, Caldwell MM, Biggs RH, eds. Climatic Impact Assessment Program (CIAP). Monograph 5, Part 1—Ultraviolet Radiation Effects, U.S. Dept. Transp., Report No. DOT-TST-75-55 (PB-247-725). Springfield, VA: Natl Tech Infor Serv, 1975:4-263 to 4-273 (Part 1, Chpt. 4, Appendix N).
46. Krizek DT. Influence of ultraviolet radiation on germination and early seedling growth. Physiol Plant 1975; 34(3):182-186.
47. Bennett JH. Photosynthesis and gas diffusion in leaves of selected crop plants exposed to ultraviolet-B radiation. J Environ Qual 1981; 10(3):271-275.

48. Lydon J, Teramura AH, Coffman CB. UV-B radiation effects on photosynthesis, growth and cannabinoid production of two *Cannabis sativa* chemotypes. Photochem Photobiol 1987; 46(2):201-206.
49. Caldwell MM, Sisson WB, Fox FM et al. Plant growth response to elevated UV irradiation under field and greenhouse conditions. In: Nachtwey DS, Caldwell MM, Biggs RH, eds. Climatic Impact Assessment Program (CIAP). Monograph 5, Part 1—Ultraviolet Radiation Effects, U.S. Dept. Transp., Report No. DOT-TST-75-55 (PB-247-725). Springfield, VA: Natl Tech Infor Serv, 1975:4-253 to 4-259 (Part 1, Chpt. 4, Appendix M).
50. Hashimoto T, Tajima M. Effects of ultraviolet irradiation on growth and pigmentation in seedlings. Plant Cell Physiol 1980; 21(8):1559-1572.
51. Dumpert K, Boscher J. Response of different crop and vegetable cultivars to UV-B irradiance: Preliminary results. In: Bauer H, Caldwell MM, Tevini M et al, eds. Biological Effects of UV-B Radiation. Munich: Bereich Projektträgerschaften, Gesellschaft für Strahlen—und Umweltforschung mbH, 1982:102-107.
52. Tevini M, Iwanzik W, Thoma U. The effects of UV-B irradiation on higher plants. In: Calkins J, ed. The Role of Solar Ultraviolet Radiation in Marine Ecosystems. New York: Plenum Pub. Corp., 1982:581-615.
53. Tevini M, Iwanzik W, Thoma U. Some effects of enhanced UV-B irradiation on the growth and composition of plants. Planta (Berl) 1981; 153(4):388-394.
54. Tevini M, Iwanzik W, Steinmüeller D et al. Effect of enhanced UV-B irradiation on growth, function and composition of plants. Proc Int Bot Congr 1981; 13:83.
55. Steinmüeller D, Tevini M. UV-B-induced effects upon cuticular waxes of cucumber, bean and barley leaves. In: Worrest RC, Caldwell MM, eds. Stratospheric Ozone Reduction, Solar Ultraviolet Radiation and Plant Life; Workshop on the Impact of Solar Ultraviolet Radiation Upon Terrestrial Ecosystems: 1. Agricultural Crops, 27-30 September 1983, Bad Windsheim, West Germany. New York: Springer-Verlag, 1986:261-269.
56. Thai VK, Garrard LA. Effects of UV-B radiation on the net photosynthesis and the rates of partial photosynthetic reactions of some crop plants. In: Nachtwey DS, Caldwell MM, Biggs RH, eds. Climatic Impact Assessment Program (CIAP). Monograph 5, Part 1—Ultraviolet Radiation Effects, U.S. Dept. Transp., Report No. DOT-TST-75-55 (PB-247-725). Springfield, VA: Natl Tech Infor Serv, 1975:4-125 to 4-145 (Part 1, Chpt. 4, Appendix F).
57. Van TK, Garrard LA. Effect of UV-B radiation on net photosynthesis of some C3 and C4 crop plants. Soil Crop Sci Soc Fla Proc 1976; 35:1-3.
58. Van TK, Garrard LA, West SH. Effects of UV-B radiation on net photosynthesis of some crop plants. Crop Sci 1976; 16(5):715-718.
59. Basiouny FM, Van TK, Biggs RH. Some morphological and bio-

chemical characteristics of C3 and C4 plants irradiated with UV-B. Physiol Plant 1978; 42(1):29-32.
60. Biggs RH, Basiouny FM. Plant growth responses to elevated UV-B irradiation under growth chamber, greenhouse and solarium conditions. In: Nachtwey DS, Caldwell MM, Biggs RH, eds. Climatic Impact Assessment Program (CIAP). Monograph 5, Part 1—Ultraviolet Radiation Effects, U.S. Dept. Transp., Report No. DOT-TST-75-55 (PB-247-725). Springfield, VA: Natl Tech Infor Serv, 1975:4-197 to 4-249 (Part 1, Chpt. 4, Appendix L).
61. Biggs RH, Sinclair TR, N'Diaye O et al. Field trials with soybeans, rice and wheat under UV-B irradiances simulating several O_3 depletion levels. In: Bauer H, Caldwell MM, Tevini M et al, eds. Biological Effects of UV-B Radiation. Munich: Bereich Projektträgerschaften, Gesellschaft für Strahlen—und Umweltforschung mbH, 1982:64-70.
62. Ambler JE, Rowland RA, Maher NK. Response of selected vegetable and agronomic crops to increased UV-B irradiation under greenhouse conditions. In: UV-B Biological and Climatic Effects Research (BACER), FY 77-78 Research Report on the Impacts of Ultraviolet-B Radiation on Biological Systems: A Study Related to Stratospheric Ozone Depletion. Final Report, Vol. I, SIRA File No.142.21, EPA-IAG-D6-0168. Washington, DC: USDA EPA, Stratospheric Impact Research and Assessment Program (SIRA), U.S. EPA, 1978.
63. Biggs RH, Webb PG. Effects of enhanced ultraviolet-B radiation on yield, and disease incidence and severity for wheat under field conditions. In: Worrest RC, Caldwell MM, eds. Stratospheric Ozone Reduction, Solar Ultraviolet Radiation and Plant Life; Workshop on the Impact of Solar Ultraviolet Radiation Upon Terrestrial Ecosystems: 1. Agricultural Crops, 27-30 September 1983, Bad Windsheim, West Germany. New York: Springer-Verlag, 1986:303-311.
64. Teramura AH. Effects of ultraviolet-B irradiances on soybean. I. Importance of photosynthetically active radiation in evaluating ultraviolet-B irradiance effects on soybean and wheat growth. Physiol Plant 1980; 48(2):333-339.
65. Webb PG. Ultraviolet-B radiation influences *Triticum aestivum* growth, productivity and microflora. Phytopathology 1982; 72(7):941.
66. Moore III FD, Burke MJ, Becwar MR. High altitude studies of natural, supplemental and deletion of UV-B on vegetables and wheat. In: UV-B Biological and Climatic Effects Research (BACER), FY 77-78 Research Report on the Impacts of Ultraviolet-B Radiation on Biological Systems: A Study Related to Stratospheric Ozone Depletion. Final Report, Vol. II, SIRA File No.142.26, EPA-IAG-D6-0168. Washington, DC: USDA EPA, Stratospheric Impact Research and Assessment Program (SIRA), U.S. EPA, 1978.
67. Gold WG, Caldwell MM. The effects of ultraviolet-B radiation on plant competition in terrestrial ecosystems. Physiol Plant 1983; 58:435-444.

68. Barnes PW, Jordan PW, Gold WG et al. Competition, morphology and canopy structure in wheat (*Triticum aestivum* L.) and wild oat (*Avena fatua* L.) exposed to enhanced ultraviolet-B radiation. Funct Ecol 1988; 2:319-330.
69. Allen LH, Vu CV, Berg III RH et al. Impact of solar UV-B radiation on crops and crop canopies. II. Effects of supplemental UV-B on growth of some agronomic crop plants. In: UV-B Biological and Climatic Effects Research (BACER), FY 77-78 Research Report on the Impacts of Ultraviolet-B Radiation on Biological Systems: A Study Related to Stratospheric Ozone Depletion. Final Report, Vol. II, SIRA File No.142.25, EPA-IAG-D6-0168. Washington, DC: USDA EPA, Stratospheric Impact Research and Assessment Program (SIRA), U.S. EPA, 1978.
70. Ambler JE, Rowland RA, Maher NK. Response of selected vegetable and agronomic crops to increased UV-B irradiation under field conditions. In: UV-B Biological and Climatic Effects Research (BACER), FY 77-78 Research Report on the Impacts of Ultraviolet-B Radiation on Biological Systems: A Study Related to Stratospheric Ozone Depletion. Final Report, Vol. I, SIRA File No.142.21, EPA-IAG-D6-0168. Washington, DC: USDA EPA, Stratospheric Impact Research and Assessment Program (SIRA), U.S. EPA, 1978.
71. Halsey LH, Kossuth SV, Biggs RH et al. Effect of ultraviolet-B radiation on sweet corn, potato and tomato under field conditions. Hort Sci 1978; 13(3 Sect. 2):359.
72. Vu CV, Allen Jr LH, Garrard LA. Effects of supplemental ultraviolet radiation (UV-B) on growth of some agronomic crop plants. Soil Crop Sci Soc Fla Proc 1979; 38:59-63.
73. Bartholic JF, Halsey LH, Garrard LA. Field trials with filters to test for effects of UV radiation on agricultural productivity. In: Nachtwey DS, Caldwell MM, Biggs RH, eds. Climatic Impact Assessment Program (CIAP), Monograph 5, Part 1—Ultraviolet Radiation Effects, U.S. Dept. Transp., Report No. DOT-TST-75-55 (PB-247-725). Springfield, VA: Natl Tech Infor Serv, 1975:4-61-4-71 (Part 1, Chpt. 4, Appendix A).
74. Tevini M, Thoma U, Iwanzik W. Effect of enhanced UV-B radiation on development and composition of plants. In: Bauer H, Caldwell MM, Tevini M et al, eds. Biological Effects of UV-B Radiation. Munich: Bereich Projektträgerschaften, Gesellschaft für Strahlen— und Umweltforschung mbH, 1982:71-82.
75. Kossuth SV, Biggs RH. Leguminosae responses to increased UV-B radiation. Plant Physiol 1979; 63(5 Suppl):107.
76. Vu CV, Allen Jr LH, Garrard LA. Effects of supplemental UV-B radiation on growth and leaf photosynthetic reactions of soybean (*Glycine max*). Physiol Plant 1981; 52(3):353-362.
77. Teramura AH, Perry MC. UV-B irradiation effects on soybean photosynthetic recovery from water stress. In: Bauer H, Caldwell MM, Tevini M et al, eds. Biological Effects of UV-B Radiation. Munich:

Bereich Projektträgerschaften, Gesellschaft für Strahlen—und Umweltforschung mbH, 1982:192-202.
78. Teramura AH, Perry MC, Lydon J et al. Effects of ultraviolet-B radiation on plants during mild water stress. III. Effects on photosynthetic recovery and growth in soybean. Physiol Plant 1984; 60(4):484-492.
79. Murali NS, Teramura AH. Effects of ultraviolet-B irradiance on soybean. VI. Influence of phosphorus nutrition on growth and flavonoid content. Physiol Plant 1985; 63(4):413-416.
80. Lydon J, Teramura AH, Summers EG. Effects of ultraviolet-B radiation on the growth and productivity of field grown soybean. In: Worrest RC, Caldwell MM, eds. Stratospheric Ozone Reduction, Solar Ultraviolet Radiation and Plant Life; Workshop on the Impact of Solar Ultraviolet Radiation Upon Terrestrial Ecosystems: 1. Agricultural Crops, 27-30 September 1983, Bad Windsheim, West Germany. New York: Springer-Verlag, 1986:313-325.
81. Murali NS, Teramura AH. Effectiveness of UV-B radiation on the growth and physiology of field-grown soybean modified by water stress. Photochem Photobiol 1986; 44(2):215-220.
82. Teramura AH, Murali NS. Intraspecific differences in growth and yield of soybean exposed to ultraviolet-B radiation under greenhouse and field conditions. Environ Exp Bot 1986; 26(1):89-95.
83. Murali NS, Teramura AH. Insensitivity of soybean photosynthesis to ultraviolet-B radiation under phosphorus deficiency. J Plant Nutr 1987; 10(5):501-516.
84. Teramura AH, Sullivan JH. Soybean growth responses to enhanced levels of ultraviolet-B radiation under greenhouse conditions. Amer J Bot 1987; 74(7):975-979.
85. Murali NS, Teramura AH, Randall SK. Response differences between two soybean cultivars with contrasting UV-B radiation sensitivities. Photochem Photobiol 1988; 48(5):653-657.
86. Teramura AH, Sullivan JH. Effects of ultraviolet-B radiation on soybean yield and seed quality: A six-year field study. Environ Pollut 1988; 53(1-4):466-468.
87. Teramura AH. The amelioration of UV-B effects on productivity by visible radiation. In: Calkins J, ed. The Role of Solar Ultraviolet Radiation in Marine Ecosystems. New York: Plenum Pub. Corp., 1982:367-382.
88. Murali NS, Teramura AH. Effects of supplemental ultraviolet-B radiation on the growth and physiology of field-grown soybean. Environ Exp Bot 1986; 26(3):233-242.
89. Krizek DT, Schaefer RL, Rowland RA. Influence of UV-B radiation on vegetative growth of *Pisum sativum* L. 'Alaska'. Hort Sci 1976; 11(3 Sect 2):306.
90. Brandle JR, Campbell WF, Sisson WB et al. Net photosynthesis, electron transport capacity and ultrastructure of *Pisum sativum* L. exposed to ultraviolet-B radiation. Plant Physiol 1977; 60:165-169.

91. Basiouny FM. Effects of UV-B irradiation on growth and development of different vegetable crops. Proc Fla State Hort Soc 1982; 95:356-359.
92. Vu CV, Allen Jr LH, Garrard LA. Effects of enhanced UV-B radiation (280-320 nm) on ribulose-1,5-biphosphate carboxylase in pea and soybeans. Environ Expt Bot 1984; 24(2):131-143.
93. Fox FM, Caldwell MM. Competitive interaction in plant populations exposed to supplementary ultraviolet-B radiation. Oecologia 1978; 36:173-190.
94. Prudot A, Basiouny FM. Absorption and translocation of some growth regulators by tomato plants growing under UV-B radiation and their effects on fruit quality and yield. Proc Fla State Hort Soc 1982; 95:374-376.
95. Nakazawa F, Tamai F, Kaneki Y. Studies on the effect of radiation of different wavelength ultraviolet rays on the growth and photosynthesis of cucumber plant. Bull Fac Agr Meiji Univ 1977; 40:7-15.
96. Krizek DT. Differential sensitivity of two cultivars of *Cucumis sativus* L. to increased UV-B irradiance. Plant Physiol 1978; 61(4 Suppl):92.
97. Krizek DT. Differential sensitivity of two cultivars of cucumber (*Cucumis sativus* L.) to increased UV-B irradiance: I. Dose-response studies. In: UV-B Biological and Climatic Effects Research (BACER), FY 77-78 Research Report on the Impacts of Ultraviolet-B Radiation on Biological Systems: A Study Related to Stratospheric Ozone Depletion. Final Report, Vol. I, SIRA File No.142.21, EPA-IAG-D6-0168. Washington, DC: USDA EPA, Stratospheric Impact Research and Assessment Program (SIRA), U.S. EPA, 1978.
98. Murali NS, Teramura AH. Intraspecific differences in *Cucumis sativus* sensitivity to ultraviolet-B radiation. Physiol Plant 1986; 68(4):673-677.
99. Tevini M, Iwanzik W. Effects of UV-B radiation on growth and development of cucumber seedlings. In: Worrest RC, Caldwell MM, eds. Stratospheric Ozone Reduction, Solar Ultraviolet Radiation and Plant Life; Workshop on the Impact of Solar Ultraviolet Radiation Upon Terrestrial Ecosystems: 1. Agricultural Crops, 27-30 September 1983, Bad Windsheim, West Germany. New York: Springer-Verlag, 1986:271-285.
100. Lipton WJ. Ultraviolet radiation as a factor in solar injury and vein tract browning of cantaloupes. J Amer Soc Hort Sci 1977; 102(1):32-36.
101. Lipton WJ, O'Grady JJ. Solar injury of "Crenshaw" muskmelons: The influence of ultraviolet radiation and of high tissue temperatures. Agr Meteorol 1980; 22(3-4):235-248.
102. Renquist AR, Hughes H, Rogoyski MK. Ultraviolet, light, and high temperature influences on simulated solar injury of red raspberry fruit. Hort Sci 1987; 22:1084.
103. Kossuth SV, Biggs RH. Sun burned blueberries. Proc Fla State Hort Soc 1978; 91:173-175.

104. Kossuth SV, Biggs RH. Ultraviolet radiation affects blueberry fruit quality. Hort Sci 1981; 14(2):145-150.
105. Krizek DT, Semeniuk P. Influence of ultraviolet radiation on the growth and development of *Browallia speciosa* Hook. Hort Sci 1974; 9(3 Sect 2):301-302.
106. Rangarajan A, Tibbitts T. The influence of radiation spectra on the development of oedema injury in ivy geraniums. Hort Sci 1988; 23:788.
107. Maekawa S, Terabun M, Nakamura N. Effects of ultraviolet and visible light on flower pigmentation of 'Ehigasa' roses. J Jap Soc Hort Sci 1980; 49(2):251-259.
108. Semeniuk P, Stewart RN. Seasonal effect of UV-B radiation on poinsettia cultivars. J Amer Soc Hort Sci 1979; 104(2):246-248.
109. Semeniuk P, Stewart RN. Comparative sensitivity of cultivars of Coleus to increased UV-B radiation. J Amer Soc Hort Sci 1979; 104(4):471-474.
110. Bogenrieder A, Klein R. Does solar UV influence the competitive relationship in higher plants? In: Calkins J, ed. The Role of Solar Ultraviolet Radiation in Marine Ecosystems. New York: Plenum Pub Corp, 1982:641-649.
111. Bogenrieder A, Douté Y. The effect of UV on photosynthesis and growth in dependence of mineral nutrition (*Lactuca sativa* L. and *Rumex alpinus* L.). In: Bauer H, Caldwell MM, Tevini M et al, eds. Biological Effects of UV-B Radiation. Munich: Bereich Projektträgerschaften, Gesellschaft für Strahlen—und Umweltforschung mbH, 1982:164-168.
112. Elawad SH, Allen Jr LH, Gascho GJ. Influence of UV-B radiation and soluble silicates on the growth and nutrient concentration of sugarcane. Soil Crop Sci Soc Fla Proc 1985; 44:134-141.
113. Inagaki N, Maekawa S, Terabun M. Effect of ultraviolet radiation on growth and photosynthetic ability of turnip (*Brassica campestris* L.). J Jap Soc Hort Sci 1986; 55(3):296-302.
114. Tevini M, Thoma U, Iwanzik W. Effects of enhanced UV-B radiation on germination, seedling growth, leaf anatomy and pigments of some crop plants. Z Pflanzenphysiol 1983; 109(5):435-448.
115. Iwanzik W. Interaction of UV-A, UV-B, and visible radiation on growth, composition and photosynthetic activity in radish seedlings. In: Worrest RC, Caldwell MM, eds. Stratospheric Ozone Reduction, Solar Ultraviolet Radiation and Plant Life; Workshop on the Impact of Solar Ultraviolet Radiation Upon Terrestrial Ecosystems: 1. Agricultural Crops, 27-30 September 1983, Bad Windsheim, West Germany. New York: Springer-Verlag, 1986:287-301.

CHAPTER 4

Mechanisms of Action

Introduction

According to Runeckles and Krupa,[1] as a prelude to discussing the consequences to terrestrial vegetation of decreased stratospheric O_3 and increased tropospheric UV-B and O_3, it is necessary to have a clear understanding of the relevant atmospheric processes that involve UV-B and O_3. Tropospheric O_3 is formed as a result of the photolysis of NO_2, as shown in Reactions 1 and 2:

$$NO_2 + h\nu \; (\lambda \leq 430 \text{ nm}) \rightarrow NO + O \qquad (1)$$

$$O + O_2 + M \rightarrow O_3 + M \qquad (2)$$

This photolysis has a peak wavelength of approximately 398 nm.[2] In contrast, the photolysis of O_3 occurs in the 290-310 nm waveband (UV-B):

$$O_3 + h\nu \; (\lambda \leq 310 \text{ nm}) \rightarrow O(^1D) + O^2 \qquad (3)$$

$$O(^1D) + H_2O \rightarrow 2HO \qquad (4)$$

The HO radical produced has been referred to as "the tropospheric vacuum cleaner"[3] because of its reactivity with a wide range of atmospheric gases including CH_4, CO and O_3.[2,4-6] The products of these subsequent reactions include peroxy radicals and hydrogen peroxide, H_2O_2.

Increases in tropospheric UV-B flux densities can only result in any net increase in local tropospheric O_3 concentrations in urban atmospheres supplied with sufficient NO, and CH_4 or CO, otherwise they lead to the increased photolytic reduction of O_3 levels.[6] Only increases in the appropriate radiation bands at wavelengths greater than those of UV-B can lead to direct elevated tropospheric O_3

Elevated Ultraviolet (UV)-B Radiation and Agriculture, by Sagar V. Krupa, Ronald N. Kickert and Hans-Jürgen Jäger.
© 1998 Springer-Verlag and Landes Bioscience.

production, provided that the supplies of NO_x and other precursors are not limiting (Reactions 1 and 3). Thus, the key concern regarding losses in stratospheric O_3 is that, of the total solar spectrum, only the penetration by UV-B will be increased.[7]

However, it should be recalled that elevated ambient O_3 (and particulate matter) levels significantly attenuate the intensity of surface UV-B.[8] Penkett[9] has suggested that surface UV-B levels in the Northern Hemisphere may be decreasing, based on the observed increases in tropospheric O_3, the negative Robertson-Berger meter trends reported by Scotto,[10] and the Brühl and Crutzen model. Indeed, Madronich et al[11] estimate that in industrialized regions the reduction of tropospheric UV-B may be as great as 10%, although it is not known whether such large attenuations may also apply to remote regions.

At the present time, losses in the stratospheric O_3 column have been coupled to the accumulation of O_3-destroying chemicals in the stratospheric clouds during the polar winter and the initiation of O_3-destroying reactions during polar sunrise.[12] With due consideration to the differences in the seasons between the two hemispheres and to the geographic differences in the pollution climate of the troposphere, it is very important to note that, in noncoastal locations, tropospheric O_3 concentrations exhibit their maximum monthly mean values during spring and summer.[13] This is particularly relevant given the predicted disproportionate role of tropospheric O_3 as a filter against solar UV-B radiation.[8] It is also important to stress that although these monthly maxima occur during spring and summer, they are averages of short-term measured concentrations. Thus, they reflect the typical average daily rise and fall in ambient O_3 concentrations that occur in air masses with active photochemistry. However, at any given location and notwithstanding the smoothed trends shown by averages, O_3 levels exhibit considerable variation from year-to-year and within a year between seasons, months, weeks, days and hours. Hence, exposures of vegetation to elevated levels of O_3 are highly random in time and are geographically patchy. In turn, this variability adds to the variability in surface level UV-B caused by changes in solar angle and reflection or scattering in the troposphere. These considerations are critical in any studies on the joint effects of UV-B and O_3 on plant species of interest. In the context of global climate change, the sequence and the way in which plants are experimentally or naturally exposed to elevated UV-B and O_3 are crucial determinants of the relevance and interpretation of the results obtained.

Krupa and Kickert[14] have described a range of possible joint exposure scenarios. Table 4.1[1] presents modified descriptions of these possibilities. Case 1 describes geographic locations where there is no predicted or observed stratospheric O_3 depletion (i.e., no increase in UV-B radiation) and no marked increase in the tropospheric O_3 concentrations. Here, we should only expect normal UV-B effects and background O_3 effects on plants.

Case 2 defines the situation at locations where there is no predicted or observed stratospheric O_3 depletion and hence no increase in UV-B, but continued upward trends in tropospheric O_3 concentrations. This case is subdivided, since in some situations (Case 2a), the surface ambient O_3 levels may be high enough to cause local adverse effects, but insufficient to increase the total tropospheric column and thereby reduce the surface UV-B flux significantly. In Case 2b, the tropospheric O_3 column is sufficient to attenuate the surface UV-B levels significantly.[1] This would lead to sub-normal UV-B levels coinciding with elevated O_3 and result in an alternation between exposures to UV-B and O_3. These cases define situations such as southern California (U.S.) and many other low to mid-latitude locations that are subjected to photochemical oxidant pollution, in which ambient O_3 is the dominant factor. It is the impact of this type of situation that air pollution–plant effects scientists have addressed for several years. However, we are unaware of any information to indicate whether the numerous published studies of the effects of O_3 would fall into Case 2a or 2b, since in no cases was UV-B flux density reported.

In those geographic areas where stratospheric O_3 depletion might occur, one might expect an increase in UV-B if spring-summer cloud conditions are not significantly increased. At locations not subjected to boundary layer O_3 concentrations significantly above the background, we might expect some plants to respond to increased UV-B (Case 3). Any interactive effect of enhanced UV-B with an atmospheric variable would probably involve increased ambient CO_2. Although we have not specifically examined the interactive effects of increased CO_2, most of the photobiology research cited in this chapter, and especially in the reports of CIAP (Climatic Impact Assessment Program, U.S.) and BACER (Biological and Climatic Effects Research, U.S.) projects in the early and mid-1970s, used this

Table 4.1. *Various possible patterns of environmental stress for vegetation with respect to O_3 and UV-B, depending upon stratospheric O_3 status and ground level O_3 pollution*[1]

Surface boundary layer status	Mid-latitudes stratospheric O_3 status	
	No O_3 depletion	O_3 depletion
Background O_3 only	[1] "Normal" UV-B plant effects with no pollution effects	[3] Enhanced UV-B plant effects only with no pollution effects
Elevated O_3 pollution	[2a] "Normal" UV-B plant effects and O_3 effects	[4a] *Enhanced* incoming UV-B might be *depleted in boundary layer* with no net effect of UV-B on plants, BUT with O_3 effects (similar to case at left)
	[2b] "Sub-normal" UV-B effects due to attenuation in boundary layer, co-occurring or intermittent with O_3 effects	[4b] Enhanced UV-B effects on plants *co-occurring or intermittent with* O_3 effects

type of situation as a frame of reference. An additional important factor that should be considered is the UV-B filtration effects of atmospheric aerosols (chapter 1). High levels of fine particle atmospheric aerosols invariably co-occur with high ambient O_3 concentrations.[2]

Mechanisms of Action of UV-B and Ozone

In many plant species, the accumulation of aromatic compounds is a general response to a variety of stresses.[15-17] Under sufficient photosynthetic photon flux, most higher plants accumulate UV-absorbing pigments (particularly flavonoids and other phenylpropanoids) in their leaves in response to elevated UV-B exposures.[18-20] Similarly, acute O_3 exposures can lead to foliar accumulation of anthocyanin,[21] caffeic acid,[22] the isoflavonoid sojagol[23] and other secondary metabolites (Fig. 4.1).

Photosystems that are capable of specifically absorbing radiation in the UV-B band mediate the induction of changes in the shikimic acid pathway.[18,24] The shikimic acid pathway leads to the production of metabolic products such as indole acetic acid (IAA, through tryptophan), and phenylpropanoids such as the flavonoids, tannins and lignin (through phenylalanine). Both IAA and phenolic compounds are oxidized by peroxidase and polyphenol oxidase (in the case of IAA by microbial laccase, Goodman et al;[25] and most recently, laccase has also been identified in higher plants during lignification, Bao et al[26]). The preference and rates of oxidation of these substrates are highly plant specific. Nevertheless, as has already been noted, in many cases UV-B exposure leads to the accumulation of flavonoids.[18] In addition to acting possibly as internal filters to UV flux (e.g., see Tevini[27]), flavonoids may also serve as defense mechanisms against insect pests and plant pathogens[28] and determine the quality of the consumed plant product.[29]

Ultraviolet-B radiation is absorbed by the entire exposed leaf surface and thus, UV-B-stimulated pigment accumulation in the epidermal tissue may offer protection against further UV-B stress.[27,30] In comparison, the primary pathway of O_3 absorption by foliage is through the stomata; the mesophyll cells are the sites of injury.[31,32] In comparison to the UV-B-induced foliar response, O_3-induced accumulation of phenolic pigments is associated with tissue senescence and necrosis or death. Differences in these plant responses to UV-B and O_3 are undoubtedly due to differences in the types of leaf cell

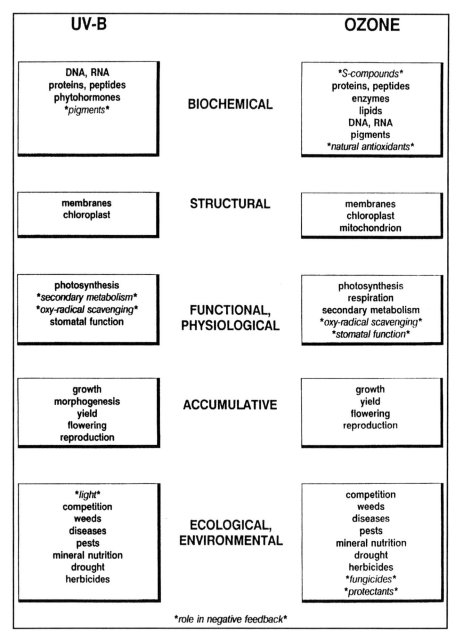

Fig. 4.1. Generalized comparison of responses of plants to UV-B and O_3 at various organizational levels. Reprinted with permission from: Runeckles VC, Krupa SV. Environ Pollut 1994; 83:191-213. © 1994 Elsevier Science Ltd.

tissue initially affected by the two forms of stress, in the rates and amounts of accumulation of the secondary metabolites and in the chemical nature of the secondary metabolites that accumulate. The concept of differing secondary metabolic responses to different forms of stress was previously suggested by Kosuge and Kimpel.[15]

In addition, both elevated UV-B[30] and O_3,[32] are known to lead to reductions of photosynthetic pigments and in some circumstances, alterations in stomatal function and hence inhibition of photosynthesis in most species. The UV-B-induced partial inhibition of photosystem II has been the subject of much research.[33] However, under field conditions there is conflicting evidence as to whether or not photosynthesis rates are significantly reduced. Again, the differences in the growth conditions between growth chamber or greenhouse studies and the field experiments, due to the differences in their photosynthetic photon flux density and possible variations in photorepair under the two sets of conditions, undoubtedly contributed to the different responses that have been reported. Other than direct effects on photosynthesis, other UV-B absorbing chromophores have not been well studied. Cinnamic acid (its *cis-trans* isomerization balance is shifted towards the *cis* form by progressively shorter wavelength UV-B radiation) has been suggested as an important UV-B receptor (M. Tevini, pers. comm.). Growth reductions due to UV-B have also been attributed to direct damage to DNA (and possibly RNA) and to the oxidation of indoleacetic acid (IAA), a plant hormone or growth regulating metabolite[20] (M. Tevini, pers. comm.).

UV-B-induced damage to DNA is usually thought to involve the formation of cyclobutane type dimers of pyrimidine bases, and several repair mechanisms, including specialized photoreactivating enzyme systems[34] have been described (see also chapter 10 in this book). In the case of O_3, Floyd et al[35] reported the detection of 8-hydroxyguanine in chloroplast DNA following the exposure of pea (*Pisum sativum* L.) and bean (*Phaseolus vulgaris* L.) plants. There is also evidence that photo-oxidation reactions leading to the production of oxy-radicals may be involved in UV-B injury, although such radicals are also known to result from irradiation at other wavelengths.[20] The involvement of such free radicals provides a further linkage with the mechanism of O_3 injury to plants and the scavenging methods that have evolved to control the levels of free radicals.[32]

Although involvement of the superoxide anion in O_3-induced injury has received the most attention, the evidence has been largely indirect and based upon changes in the levels of the ubiquitous scavenging enzyme superoxide dismutase (SOD). However, Runeckles and Vaartnou[36] have been able to detect an electron paramagnetic resonance (EPR) signal in intact, attached leaves of *Poa pratensis* L. and *Lolium perenne* L. exposed to O_3 that has the spectral characteristics of the O_2^- radical. The signal could only be detected during exposures that presumably were sufficient to overcome the abilities of tissues to scavenge the radical. Attempts to detect O_2^- and other radicals in situ by means of their adducts with various spin labels have been unsuccessful, because of the toxicity of the spin labels.[36] However, Melhorn[37] provided evidence for an unidentified radical in pea and bean leaves treated with O_3 by means of surface application of the spin trap $-t$-butyl-α-phenylnitrone. Most recently, Runeckles and Vaartnou[38] used electron paramagnetic resonance spectrometry and identified light mediated (chloroplast) O_2^- production in Kentucky blue grass (*Poa pratensis* L.), ryegrass (*Lolium perenne* L.) and radish (*Raphanus sativus* L.) in response to O_3 exposures.

In the context of plant disease, the effects of O_3 on the host, pathogen or both may also involve the participation of the superoxide anion radical (O_2^-) and hydrogen peroxide. Rapid increases in the levels of O_2^- were detected in the hypersensitive reaction induced by infection of Burley 21 tobacco (*Nicotiana tabacum* L.) with *Pseudomonas syringae* pv. *syringae*.[39] In addition, Apostol et al[40] reported a rapid burst of H_2O_2 released by cultured plant cells upon treatment with elicitors of defense responses. Hence O_2^- and H_2O_2 may well be intimately involved in the interactive responses of O_3, host and the pathogen, with the diversity of responses reflecting the degree to which the effects of the pathogen are limited to the sites of infection or spread to adjacent cells and modifying their scavenging systems.

Hence, in spite of their inherent differences as stress factors, at the mechanistic level UV-B and O_3 appear to have some features in common. As a result, if SOD plays a key role in controlling O_2^-, the combination of UV-B and O_3 stress may lead to O_2^- levels that exceed the control capacity of SOD, with a consequent increase in injury response.

Eco-Physiological Adaptations and Evolution

There is evidence that certain chromophores may serve as specific UV-B receptors.[18] According to Caldwell et al,[33] when a "UV-B receptor" is involved, the induction of the reaction is modulated rather than quantal (all or nothing) in nature. Thus, although plants have been exposed to some UV-B radiation, they can further respond to additional UV-B flux. Flint et al[41] observed that a UV-B supplement can elicit further flavonoid accumulation even when the plants were grown under full solar spectrum. Thus, modulated response can occur under the present and potential enhanced UV-B conditions, affording plants an opportunity for stress repair or avoidance. However, this capability will vary at the generic, species, varietal or cultivar levels and will be important in competition and species survival in mixed populations (see also chapter 9 in this book).

Both short- and long-term adaptations have been reported in response to exposure to O_3. Extending the early observations of Walmsley et al[42] that leaves of radish (*Raphanus sativus* L.) plants exposed to O_3 became progressively able to withstand the adverse effects of increased O_3 levels, the studies of Held et al[43] led Runeckles and Krupa[1] to suggest that such acclimation has a developmental rather than a physiological mechanism. At the genetic level, Heagle et al[44] have shown that during prolonged exposures to O_3 (two growing seasons), selection pressure on clover (*Trifolium* spp.) for resistance to O_3 stress led to the isolation of a high proportion of resistance clones. Therefore, this result can be extended to an agricultural crop, together with the previous evidence offered for the genetic adaptation to O_3 by a tree species, aspen (*Populus tremuloides* Michaux).[45]

A summary of the comparative responses of plants to UV-B and O_3 at various organizational levels is presented in Figure 4.1. References to the specific responses listed are to be found in Worrest and Caldwell,[7] Tevini and Teramura,[30] Strid et al,[34] Bornman and Teramura[46] and Tevini[47] for UV-B, and Runeckles and Chevone[32] and Alscher and Wellburn[48] for O_3. Some of the responses listed are well documented; however, in some cases, the evidence is conflicting and others are largely inferential. Although several responses are common to both stresses, certain responses have been highlighted because of their potential defensive or protective functions. It is worth noting the demonstrated effects of O_3 on mitochondria and respiration and the probable importance of natural antioxidants such as ascorbate and glutathione in providing protection. Certain fungicides and other compounds may also offer some protection against O_3. On the other hand, light intensity plays an important role in photorepair of UV-B-induced injury and stimulated secondary metabolism leading to UV-absorbing compounds such as the flavonoids.

References

1. Runeckles VC, Krupa SV. The impact of UV-B and ozone on terrestrial vegetation. Environ Pollut 1994; 83:191-213.
2. Finlayson-Pitts BJ, Pitts Jr JN. Atmospheric Chemistry: Fundamentals and Experimental Techniques. New York: John Wiley & Sons, 1986.
3. Graedel TE. Chemical Compounds in the Atmosphere. New York: Academic Press, 1978.
4. Calvert JG, ed. SO_2, NO and NO_2 Oxidation Mechanisms: Atmospheric Considerations, Acid Precipitation Series. Teasley JI, series ed. Boston: Butterworth, 1984:1-62.
5. Demerjian KL. Atmospheric chemistry of ozone and nitrogen oxides. In: Legge AH, Krupa SV, eds. Air Pollutants and Their Effects on the Terrestrial Ecosystem. New York: John Wiley & Sons, 1986:105-127.
6. Thompson AM. The oxidizing capacity of the earth's atmosphere: Probably past and future changes. Science 1992; 256:1157-1165.
7. Worrest RC, Caldwell MM, eds. Stratospheric Ozone Reduction, Solar Ultraviolet Radiation and Plant Life; NATO Advanced Research Workshop on the Impact of Solar Ultraviolet Radiation Upon Terrestrial Ecosystems: 1. Agricultural Crops; 27-30 September 1983, Bad Windsheim, West Germany. NATO ASI Series, Series G, Ecological Sciences, Vol. 8. Berlin: Springer-Verlag, 1986.
8. Brühl C, Crutzen PJ. On the disproportionate role of tropospheric ozone as a filter against solar UV-B radiation. Geophys Res Lett 1989; 16:703-706.
9. Penkett SA. Ultraviolet levels down not up. Nature 1989; 341:283-284.
10. Scotto J, Cotton G, Urbach F et al. Biologically effective ultraviolet radiation: Surface measurements in the United States, 1974 to 1985. Science 1988; 239:762-764.
11. Madronich S, Björn LO, Ilyas M et al. Changes in biologically active ultraviolet radiation reaching the earth's surface. In: Environmental Effects of Ozone Depletion: 1991 Update. Nairobi, Kenya: United Nations Environment Programme, 1991:1-13.
12. Anderson JG, Toohey DW, Brune WH. Free radicals within the Antarctic vortex: The role of CRCs in Antarctic ozone loss. Science 1991; 251:39-46.
13. Logan JA. Tropospheric ozone: Seasonal behavior, trends, and anthropogenic influence. J Geophys Res 1985; 90D6(10):463-482.

14. Krupa SV, Kickert RN. The Greenhouse Effect: Impacts of ultraviolet-B (UV-B) radiation, carbon dioxide (CO_2), and ozone (O_3) on vegetation. Environ Pollut 1989; 61(4):263-392.
15. Kosuge T, Kimpel JA. Energy use and metabolic regulation in plant-pathogen interactions. In: Ayres PG, ed. Effects of Disease on the Physiology of the Growing Plant. Cambridge, England: Cambridge University Press, 1981:29-45.
16. Friend J. Alterations in secondary metabolism. In: Ayres PG, ed. Effects of Disease on the Physiology of the Growing Plant. Cambridge, England: Cambridge University Press, 1981.
17. Shirley BW. Flavonoid biosynthesis: 'new' functions for an 'old' pathway. Trends Plant Sci 1996; 1(11): 377-388.
18. Wellmann E. UV radiation: Definitions, characteristics and general effects. In: Shropshire W, Mohr H, eds. Encyclopedia of Plant Physiology, New Series, Photomorphogenesis. Berlin: Springer-Verlag, 1983:16B:745-756.
19. Tevini M, Iwanzik W, Teramura AH. Effects of UV-B radiation on plants during mild water stress. II. Effects on growth, protein and flavonoid content. Z Pflanzenphysiol 1983; 110:459-467.
20. Beggs CJ, Schneider-Ziebert U, Wellmann E. UV-B radiation and adaptive mechanisms in plants. In: Worrest RC, Caldwell MM, eds. Stratospheric Ozone Reduction, Solar Ultraviolet Radiation and Plant Life; Workshop on the Impact of Solar Ultraviolet Radiation Upon Terrestrial Ecosystems: 1. Agricultural Crops, 27-30 September 1983, Bad Windsheim, West Germany. New York: Springer-Verlag, 1986:235-250.
21. Koukol J, Dugger Jr WM. Anthocyanin formation as a response to ozone and smog treatment of *Rumex crispus* L. Plant Physiol 1967; 42:1023-1024.
22. Howell RK. Influence of air pollution on quantities of caffeic acid isolated from leaves of *Phaseolus vulgaris*. Phytopathology 1970; 60:1626-1629.
23. Keen NT, Taylor OC. Ozone injury in soybean: Isoflavonoid accumulation is related to necrosis. Plant Physiol 1975; 55:731-733.
24. Hahlbrock K, Grisebach H. Enzymic controls in the biosynthesis of lignin and flavonoids. Annu Rev Plant Physiol 1979; 30:105-130.
25. Goodman RN, Király A, Wood KR. The Biochemistry and Physiology of Plant Disease. Columbia: University of Missouri Press, 1986.
26. Bao W, O'Malley DM, Whetten R et al. A laccase associated with lignification in loblolly pine xylem. Science 1993; 260:672-674.
27. Tevini M, Braun J, Fieser G. The protective function of the epidermal layer of rye seedlings against ultraviolet-B radiation. Photochem Photobiol 1991; 53:329-333.
28. Harborne JB, ed. The Flavonoids: Advances in Research since 1980. London: Chapman & Hall, 1988.
29. Hansen RG, Wyse BW, Sorenson AW. Nutritional Quality Index of Foods. Westport, CO: AVI Publ. Co., 1979.

30. Tevini M, Teramura AH. UV-B effects on terrestrial plants. Photochem Photobiol 1989; 50(4):479-487.
31. Runeckles VC. Uptake of ozone by vegetation. In: Lefohn AS, ed. Surface Level Ozone Exposures and Their Effects on Vegetation. Chelsea, MI: Lewis Publishers Inc., 1992:157-188.
32. Runeckles VC, Chevone BI. Crop responses to ozone. In: Lefohn AS, ed. Surface Level Ozone Exposures and Their Effects on Vegetation. Chelsea, MI: Lewis Publishers Inc., 1992.
33. Caldwell MM, Teramura AH, Tevini M. The changing solar ultraviolet climate and the ecological consequences for higher plants. Trends Ecol Evolut 1989; 4:363-367.
34. Strid Å, Soon Chow W, Anderson JM. UV-B damage and protection at the molecular level in plants. Photosyn Res 1994; 39:475-489.
35. Floyd RA, West MS, Hogsett WE et al. Increased 8-hydroxyguanine content of chloroplast DNA from ozone-treated plants. Plant Physiol 1989; 91:644-647.
36. Runeckles VC, Vaartnou M. Observations on the in situ detection of free radicals in leaves using electron paramagnetic resonance spectrometry. Can J Bot 1992; 70:192-199.
37. Melhorn H, Tabner BJ, Wellburn AH. Electron spin resonance evidence for the formation of free radicals in plants exposed to ozone. Physiol Plant 1990; 79:377-383.
38. Runeckles VC, Vaartnou M. EPR evidence for superoxide anion formation in leaves during exposure to low levels of ozone. Plant, Cell Environ 1997; 20:306-314.
39. Adam A, Farkas T, Somlyai G et al. Consequence of O_2^- generation during a bacterially induced hypersensitive reaction in tobacco: Deterioration of membrane lipids. Physiol Mol Plant Pathol 1989; 34:13-26.
40. Apostol I, Heinstein PF, Low PS. Rapid stimulation of an oxidative burst during elicitation of cultured plant cells. Plant Physiol 1989; 90:109-116.
41. Flint SD, Jordan PW, Caldwell MM. Plant protective response to enhanced UV-B radiation under field conditions: Leaf optical properties and photosynthesis. Photochem Photobiol 1985; 41(1):5-99.
42. Walmsley L, Ashmore MR, Bell JNB. Adaptation of radish *Raphanus sativus* L. in response to continuous exposure to ozone. Environ Pollut (Ser. A) 1980; 23:165-177.
43. Held AA, Mooney HA, Gorham JN. Acclimation to ozone stress in radish: Leaf demography and photosynthesis. New Phytol 1991; 118:417-423.
44. Heagle AS, McLaughlin MR, Miller JE et al. Adaptation of a white clover population to ozone stress. New Phytol 1991; 119:61-66.
45. Berrang P, Karnosky DF, Bennett JP. Natural selection for ozone tolerance in *Populus tremuloides*: Field verification. Can J Forest Res 1989; 19:519-522.

46. Bornman JF, Teramura AH. Effects of ultraviolet-B radiation on terrestrial plants. In: Young AR, Moan J, Olof L et al, eds. Environmental UV Photobiology. New York: Plenum Press, 1993:427-471.
47. Tevini M, ed. UV-B Radiation and Ozone Depletion: Effects on Humans, Animals, Plants, Microorganisms and Materials. Boca Raton, FL: Lewis Publishers, 1993.
48. Alscher RG, Wellburn AR, eds. Plant Responses to the Gaseous Environment. London: Chapman & Hall, 1994.

CHAPTER 5

Crop-Weed Interactions

Introduction

If computations are made on the basis of the 1990 farm income of $80.4 billion in the U.S., then according to the estimates of Hale,[1] in the context of crop losses due to weeds, diseases, insects, and tropospheric ozone, weeds cause about 6% of the lost income. Changes in ground level UV-B flux during the growing season have the potential for altering the amount of crop loss from weed competition. This assumes relationships between a desired crop competing for growth resources with one or more undesired weed species. However, there are circumstances where weeds either provide a beneficial effect for crop growth or themselves are harvested for useful commodities.[2] We often tend to think in terms of a competitive relationship between crops and weeds, where the crop, or at least the farmer is negatively impacted by the undesirable plant interactions. However, sometimes there can be mutually beneficial interactions between a crop and its weeds.[2]

In addition to the structurally based inter-species spatial competition, another process of biological interaction between some crops and weeds occurs through allelopathy. Published literature on the effects of UV-B radiation on allelopathy between crops and weeds appears to be lacking, therefore in the following sections attention is directed to the spatial competition only.

Recent studies on changes in rural weeds over the past 40 years (1950 to 1990) in Central Europe have concluded that there has been a general decrease in weed species as a result of changes in land use systems.[3] However, an increase in: (a) herbicide-tolerant weeds, (b) certain broad-leafed (dicot) and grass (monocot) weeds, and (c) the spread of new weed species, have occurred over the same period.[4]

Elevated Ultraviolet (UV)-B Radiation and Agriculture, by Sagar V. Krupa, Ronald N. Kickert and Hans-Jürgen Jäger.
© 1998 Springer-Verlag and Landes Bioscience.

It might be assumed that the use of chemical herbicides can overcome weed problems in cropland agriculture, but this view may represent a bias. If for no other reason than mere economics, this view cannot be regarded as applicable around the world in all societies. Even in the developed countries, with the increased emphasis on environmental protection, greater emphasis is being placed on integrated pest management (IPM).

The Nature of Competition Between Two Plant Species

Whether they are all the same species, or represented by two or more species in a mixture, individual plants compete with each other above-ground for light (photosynthetically active radiation), and below-ground for water and nutrients. Aggressive above-ground competition for light can be seen through the growth patterns of stem and/or foliage, although other organs can also be affected. The few studies that have been published to date on the effects of elevated UV-B on the balance of competition between plant species, have essentially been directed to above-ground competition.[5] In this context, very little is known about possible changes in carbon partitioning between above- and below-ground plant organs.

Without affecting total plant growth, changes in UV-B exposure can cause changes in the structure or morphology of plants competing for needed resources in mixed populations such as crops and weeds. The extent to which carbon assimilation by the canopy and subsequent allocation among the organs are affected by photomorphogenetic processes, is an important question which requires further investigation.[6]

Below-ground competition for water and/or nutrients is exerted by root growth and expansion through the soil, although, through this process other organs can also be affected. In a population, if almost all plants are of the same species (e.g., 98 versus 2%), then intra-species competition can occur between individuals within the 98%, especially if too many plants are crowded into a small area, i.e., a density-effect, not described here. On the other hand, this chapter addresses the competition between two plant species such as a crop and a weed, when the proportion of the two species in the population is something in the order of 70 versus 30%, or 80 versus 20%. It is known that the tendency to display a reduction in growth can be different between species, depending upon whether an intra-species density effect is at work, or whether an inter-species competition is actively in progress.[5]

In an agricultural setting, the net effect of successful weed competition in the harvest of a desired crop can be quantitative or qualitative. There can be a quantitative decline in the harvested portion of the crop, because of decreased crop growth resulting from the competition with the weed species. In contrast, there can be a qualitative decline in the homogeneity of the harvested portion of the crop, if the mixture is not purely composed of the crop but also contaminated with weeds. This is especially relevant in grain harvesting, but not usually where the harvested part is stem, leaf, fruit, roots or tubers. For these latter types of commodities, quantitative growth decline in the harvested crop part is the most likely effect of successful weed competition. This can even be to the extent of reducing the growth of the entire crop population at the early seedling stage.

The Database on Crop-Weed Competition

A question that should be addressed is: What is the effect of a change in ambient plant effective UV-B irradiance during the growing season on the balance of competition between crop and weed species? At the present time there are no good compilations of the levels of ambient UV-B exposures that cause various levels of growth responses in a crop or in a weed species. To further complicate matters, the balance of competition between crops and weeds is also affected by the availability of resources required for growth, such as light, heat, moisture, nutrients, as well as from air pollutants such as ground-level ozone. This subject of multiple stress factors is addressed in chapter 7.

UV-B irradiance can have a negative effect on plant growth, but different plant species, and cultivars within a species have different degrees of sensitivity.[7,8] In a general sense, a competition effect is shown in Figure 5.1. At a low level of plant effective UV-B exposure, the relative growth rates of neither the crop nor the weed species might be affected (time T1) (Fig. 5.1). If the effective UV-B exposure significantly increases, at a later time (T2) the relative growth rate of the crop might decrease, more so than that of its principal weed species. Under those conditions one could expect a need for more weed control (added expense) or suffer increased crop losses.

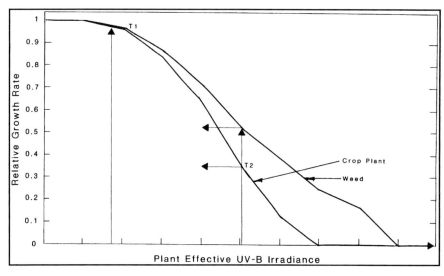

Fig. 5.1. Schematic diagram of one possible pattern of the relative growth rate of a dicot, crop plant and monocot, weed under increasing levels of plant effective UV-B irradiance.

Major Crops and Their Principal Weeds

Ideally, the principles and techniques in molecular genetics would have identified biological markers that would allow rapid detection and compilation of the sensitivity of various crop cultivars and weed species to increased UV-B irradiance at the natural levels of global (total) solar radiation. But, our knowledge is not yet at that stage and therefore, to estimate what changes in plant competition might occur under future climatic changes, if ambient UV-B levels were to increase, it is necessary to examine other approximate guidelines.

In order for UV-B radiation to initiate plant responses, it must penetrate the epidermis and into the interior mesophyll tissue. Day et al[9] presented results that might be useful for judging the qualitative change in competitive status between crops and their major weeds under increased UV-B. Epidermal transmittance is proportional to the level of incident radiation, minus what is reflected by the leaf surface and what is absorbed by the epidermis itself. Day et al[9] concluded that the least transmittance was in conifers (C), increasing in woody dicots (WD), and monocots (M), with greatest transmittance in herbaceous dicots (HD) (c < WD < M < HD). Many of the vegetable crops are herbaceous dicots, growing with weeds that are either herbaceous dicots or monocots. The grain crops are monocots competing with weeds that are either herbaceous dicots or are monocots.

The salt marsh dicot, *Aster tripolium* L. showed greater reduction in growth and photosynthesis than the corresponding monocot grass *Spartina anglica* C. Hubb.[10] However, the results were obtained from a glasshouse experiment in which UV-B exposure was simulated as a constant square wave, 9 hours per day (1.34 W m^{-2}) for four weeks. Plant response to a constant square wave exposure most likely will be different from an exposure to ambient sinusoidal, diurnal UV-B regime.[11]

A recent outdoor pot experiment by Cann[12] with two major weed species, *Amaranthus retroflexus* L. (a C4 dicot) and *Chenopodium album* L. (a C3 dicot) showed slight increase in growth rates with low UV-B, implying decreased growth with increase in UV-B. While these results are interesting, the two weeds were not grown as mixtures with crops, against which they normally compete for resources. As a result, nothing can be stated in this case about the consequences for inter-species competition under increased ambient UV-B.

Crops and their corresponding weeds represent combinations of six patterns: (1) monocot crops mainly with dicot weeds, (2) dicot crops principally with monocot weeds, (3) monocot crops and weeds, (4) dicot crops and weeds, (5) woody dicot crops with herbaceous dicot weeds, or (6) woody dicot crops with monocot weeds. There is evidence that various growth and developmental characteristics of dicots are more sensitive, and likely to be negatively impacted under increased UV-B, compared to monocots which possess greater resilience.[13] By contrast, another study showed that monocots have a tendency to be more negatively impacted by increased UV-B in mixtures with dicots.[14] That study revealed that "...changes in competitive balance resulting from increased UV-B might be expected more frequently when monocots are involved in mixtures, rather than mixtures of only dicots."[14] A study in which both the crop and the weed were monocots, in a competition for light between wheat (*Triticum aestivum* L.) and the weed wild oat grass (*Avena fatua* L.), the competitive advantage of the weed was lost under an additional dose of UV-B equivalent to a 20% loss in stratospheric ozone.[15] The authors hypothesized that grasses are morphologically more responsive to UV-B than dicots. They further suggested that mixtures of monocots (crop and weeds) might be more susceptible to changes in competition under increased UV-B, than mixtures of dicots.

The evidence from various studies is mixed on what the sequence of sensitivity to UV-B really is, but the scheme from Day et al[9] does present an approach to estimate the possible consequences to crops and their major weeds, depending upon which species are herbaceous dicots, monocots or woody dicots. With this as a rough guideline, one can estimate the future status of competition between crops and their weeds on an international or global scale.

Crops and Weeds Under Increased UV-B Radiation and/or Climate Warming

By continents and regions, the principal weeds[16,17] competing with major dicot and monocot crops are shown in Tables 5.1 and 5.2. Crops that are considered to be sensitive or responsive to exposures to increased CO_2, UV-B, or O_3[7] are indicated by shading.

The transition market countries, including Eastern Europe, lead the world in potato (*Solanum tuberosum* L.) production (chapter 8). If future changes in climate lead to increased ambient plant effective UV-B irradiance during the spring or summer, the competitive balance between potato and its principal weeds might not be affected, since all of them are dicots. However, this conclusion depends on whether the potato cultivars that are grown are themselves sensitive to UV-B. In addition, *Stellaria media* (L.) Cirillo (a weed) might exhibit an increase in the number of life cycles per year and a consequent increase in their seed distribution in the soil with increased temperatures.[19]

Western Europe leads the world in sugar beet (*Beta vulgaris* L.) production. With sugar beets, the annual grass weed, *Echinochloa crus-galli* (L.) Pal., could be expected to decline in its competitive ability, although it has been estimated to display earlier and increased germination from elevated CO_2-induced higher temperatures.[19]

Developing Africa leads the world in rangeland pasture. Pastures and meadows used for livestock grazing could see less competition from the weedy dicots, *Rumex crispus* L. and *Capsella bursa-pastoris* Medikus, under any increased ambient UV-B, although there is a projected increase in the number of life cycles of *C. bursa-pastoris* due to elevated CO_2-induced increase in temperature.[19]

From a rare field study, it was found that in comparing ambient UV-B growth responses to those under enhanced UV-B levels, the balance of competition was switched from *Geum macrophyllum* L. (a weed) to the crop Kentucky bluegrass, *Poa pratensis* L.[20] The same study also showed the balance of competition switched from *Amaranthus retroflexus* (a weed) to the alfalfa crop (*Medicago sativa* L.), with increased UV-B levels. Alfalfa is a widely prominent crop in North America (Table 5.1) and therefore, such results are of interest in that part of the world.

Grape (*Vitis* spp.) vineyards, and woody dicots, with their world production being dominated by developing Asia, and the transition market countries of Eastern Europe, might have less problems with two of the principal grass weeds, *Cynodon dactylon* (L.) Pers. and *Sorghum halepense* (L.) Pers. However, the balance could tip either way in vineyards because *C. dactylon*, with a warmer climate, is projected to extend northward into new areas from which a cooler climate has so far excluded it.[19] Here, *S. halepense* has been estimated to possibly show earlier and increased germination in a warmer climate.

Crops: Possible Winners and Losers

In summary, with regard to only competitive balance between a crop and its principal weed species and based on only very general patterns that require much more research to verify, the crops that might be considered to become winners, if increased plant effective UV-B levels occur in the future are: sugarcane (*Saccharum officinarum* L.), sorghum (*Sorghum vulgare* Moench), coffee (*Coffea arabica* L.), papaya (*Carica papaya* L.), and oranges (*Citrus sinensis* (L.) Osbeck) in Latin America and the Caribbean, vineyards in Eastern European transition market countries, and developing Asia, pasture rangelands in developing Africa, barley (*Hordeum vulgare* L.) and olive (*Olea europaea* L.) in Western Europe, rye (*Secale cereale* L.) and oats (*Avena sativa* L.) in Eastern European transition market countries, and wheat, coconut (*Cocos nucifera* L.) and oil palm (*Elaeis guineensis* Jacq.) in developing Asia. A possible decrease in the competitive ability of present major weed species might simply lead to other currently minor weed species becoming more prominent. The crop loser might be cotton (*Gossypium hirsutum* L.) in developing Asia.

There are a large number of crops for which it is uncertain whether the crop or its major weed species would be favored with a higher ground level UV-B in the future (Table 5.3). With the exception

Table 5.1 Major dicotyledonous crops of world regions with sensitivity to all three factors, CO_2 + UV-B + O_3 (shaded), and major weeds likely to produce plant competition; annual/perennial and forb/grass life forms are for the crop and do not necessarily pertain to the associated weed species[17]

	North America	Latin Am. & Caribb.	Developing Africa	Western Europe	East. Eur. Transition Countries	Developing Asia	Oceania	Major weeds
Soybean		—	—	—	—	—	—	*Acanthospermum hispidum, Amaranthus retroflexus, Cassia tora, Datura stramonium, Hibiscus trionum, Rumex acetosella, Setaria glauca*
Alfalfa		—	—	—	—	—	—	*Amaranthus retroflexus, Brassica kaber, Chenopodium murale, Conyza canadensis, Cuscuta campestris, Daucus carota, Raphanus raphanistrum, Setaria glauca, Sonchus arvensis*
Oranges	—		—	—	—	—	—	*Stellaria media*
Plantains	—		—	—	—	—	—	Not available
Coffee	—		—	—	—	—	—	*Boerhavia erecta, Drymaria cordata, Momordica charantia, Phyllanthus niruri, Sida rhombifolia*
Papaya	—		—	—	—	—	—	*Amaranthus viridis, Asclepias curassavica*

Crop-Weed Interactions

Crop						Weeds
Cassava	—	—	—	—	—	Acanthospermum hispidum, Datura stramonium, Euphorbia heterophylla, Synedrella nodiflora, Tridax procumbens
Clover	—	—	—	—	Clover	Africa: Not available; Oceania: Carduus pycnocephalus, Orobanche minor
Yams	—	—	—	—	—	Not available
Taro	—	—	—	—	—	Amaranthus viridis
—	Sugar beets	—	—	—	—	Amaranthus retroflexus, Chenopodium album, Echinochloa crus-galli, Sinapis arvensis, Solanum nigrum
—	Olives	—	—	—	—	Portulaca oleracea
—	Peas	—	—	—	—	Alopecurus myosuroides, Brassica kaber, Matricaria chamomilla, Papaver rhoeas, Poa annua, Polygonum persicaria
—	Peaches/nectarines	—	—	—	—	Not available
—	—	Grape	—	—	—	Cynodon dactylon, Sorghum halepense
—	—	Potatoes	—	—	—	Chenopodium album, Galinsoga parviflora, Stellaria media
—	—	Sunflowers	—	—	—	Hungary, Yugoslavia: Amaranthus retroflexus; Rumania: Polygonum aviculare, Yugoslavia: Orobanche ramosa, Former Soviet Union: Brassica kaber, Cardaria draba, Chondrilla juncea

cont'd.

	North America	Latin Am. & Caribb.	Developing Africa	Western Europe	East. Eur. Transition Countries	Developing Asia	Oceania	Major weeds
	—	—	—	—	—	Sweet potatoes	—	*Borreria alata, Chenopodium ambrosioides, Physalis angulata, Polygonum hydropiper, Trianthema portulacastrum*
	—	—	—	—	—	Oil palm fruit	—	*Borreria alata, Emilia sonchifolia, Melastoma malabathricum, Momordica charantia, Passiflora foetida*
	—	—	—	—	—	Coconuts	—	*Cassia tora, Cyperus brevifolius, Melastoma malabathricum, Passiflora foetida, Stachytarpheta jamaicensis*
	—	—	—	—	—	Tomatoes	—	*Chenopodium album, Cyperus rotundus, Galinsoga parviflora, Polygonum convolvulus, Portulaca oleracea*
	—	—	—	—	—	Cotton	—	*Cynodon dactylon, Sorghum halepense*
	—	—	—	—	—	Cabbage	—	*Trianthema portulacastrum*
	—	—	—	—	—	Bananas	—	*Conyza canadensis, Ipomoea aquatica*
	—	—	—	—	—	Apples	—	*Agropyron repens, Portulaca oleracea*
	—	—	—	—	—	Groundnuts	—	*Acanthospermum hispidum, Amaranthus retroflexus, Borreria alata, Digitaria longiflora, Phyllanthus niruri, Setaria glauca, Trianthema portulacastrum, Tridax procumbens, Vernonia cinerea*

Table 5.2. Major monocotyledonous crops of world regions with sensitivity to all three factors, $CO_2 + UV\text{-}B + O_3$ (shaded), and major weeds likely to produce plant competition; annual/perennial and forb/grass life forms are for the crop and do not necessarily pertain to the associated weed species[17,18]

North America	Latin Am. & Caribb.	Developing Africa	Western Europe	East. Eur. Transition Countries	Developing Asia	Oceania	Major weeds
Maize	—	—	—	—	—	—	Cynodon dactylon, Digitaria sanguinalis, Echinochloa colonum, E. crus-galli, Sorghum halepense
—	Sugarcane	—	—	—	—	—	Asclepias curassavica, Emilia sonchifolia, Ipomoea triloba, Momordica charantia, Passiflora foetida, Phyllanthus niruri, Sida rhombifolia
Sorghum	Sorghum	Sorghum	—	—	—	—	Caribbean: Brassica campestris, Euphorbia heterophylla, Sida rhombifolia, Trianthema portulacastrum; Africa: Corchorus olitorius, Digitaria velutina, Nicandra physalodes, Phyllanthus niruri, Trianthema portulacastrum
—	—	Range pasture	—	—	—	—	Bromus spp., Capsella bursa-pastoris, Rumex crispus

Crop				Weeds
Barley	—	—	—	Alopecurus myosuroides, Brassica kaber, Fumaria officinalis, Matricaria chamomilla, Papaver rhoeas, Polygonum aviculare, P. hydropiper, Raphanus raphanistrum
Triticale	—	—	—	Agropyron repens, Alopecurus myosuroides, Apera spica venti, Stellaria media, Veronica spp.
Rye	—	—	—	Poland: Papaver rhoeas; Yugoslavia: Raphanus raphanistrum; Former Soviet Union: Polygonum aviculare
Oats	—	—	—	Bulgaria: Fumaria officinalis; Lithuania: Rumex acetosella; Rumania: Polygonum hydropiper; Former Soviet Union: Brassica kaber, Cardaria draba, Chondrilla juncea, Raphanus raphanistrum
Wheat	—	—	—	Avena fatua, Polygonum convolvulus, Stellaria media
Rice	—	—	—	Cyperus difformis, C. rotundus, Echinochloa crus-galli

Table 5.3. *Crop-weed matrix among dicots and monocots with possible competitive change under increased UV-B radiation*[a]

Major weeds	Dicot	Crop type Monocot	Woody dicot
Herbaceous dicot	**Uncertain favor** Alfalfa, soybean, cassava, taro, clover, sugar beet, pea, potato, sunflower, cabbage, banana, tomato, sweet potato, groundnuts	**Crop favored** Sugarcane, sorghum, pasture, barley, rye, oat, wheat	**Crop favored** Coffee, papaya, oranges, olive, coconut, oil palm
Monocot	**Weed favored** Cotton	**Uncertain favor** Maize, wheat, rice	**Crop favored** Grape

[a]Refer to the text and Tables 5.1 and 5.2. Weeds of apple orchards have a mixed pattern and cannot be classified.

of Latin America and the Caribbean countries, all of the major land masses considered here have at least one crop-weed combination for which the future is uncertain, but developing Asia has the most.

The outlook regarding the effects of elevated UV-B can be changed if one considers possible direct and indirect (heating) effects from increased ambient CO_2 in the atmosphere on the competitive balance between crops and their weeds. In that case, the crop losers might be sugar beets, the grain crops, pastures and vineyards.

In the transition market countries, if future climate changes lead to increased ambient plant effective UV-B irradiance during the spring or summer, the competitive balance between potatoes and its principal weeds, *Chenopodium album*, *Stellaria media*, and *Galinsoga parviflora* Cav., might not be affected (all are dicots). But again, this depends on whether the potato cultivars that are grown are sensitive to increased ambient plant effective fluxes of UV-B. A small amount of evidence shows that the weed *C. album* might be favored. Such evidence showed a 9.4% increase in shoot dry weight of *C. album* when grown for 36 days as a pure stand in a glasshouse under an estimated plant effective UV-B irradiance of 2.67 W m^{-2} at Logan, Utah.[14] This was judged to be equivalent to a 20% depletion in the ozone column on 1 June, for clear skies at the study location. However, this study was not in an ambient environment and the weed was not grown as a mixture with a crop. Nevertheless, the competitive balance could also be changed in favor of the weed, if ambient CO_2 levels continued to increase. *Chenopodium album* is known to increase its seed production and seed weight when grown under elevated levels of CO_2.[19] Similarly, the increase in seed capital and life cycles per year for *S. media* (a weed) in a lengthened growing season resulting from higher temperatures has been mentioned previously.[19]

Western Europe leads the world in sugar beet production, but the crop also competes with a number of principal weeds (Table 5.1). The grass weed *Echinochloa crus-galli* could suffer relatively less from increased UV-B in competition with the crop, and increased seed germination has been estimated for the weed under higher temperatures.[19] Increased CO_2-stimulated seed production for *Chenopodium album* has been mentioned earlier, and this response has also been ascribed as being likely for *Amaranthus retroflexus* that competes with sugar beets.[19] However, *A. retroflexus* showed a 9.5% increase in shoot dry weight only when grown for 33 days as a pure stand in a glasshouse under an estimated plant effective UV-B irradiance of

2.67 W m^{-2}.[14] For the other principal weed *Solanum nigrum* L. (Table 5.1), earlier emergence and enhanced seed production has been estimated under climate warming.[19]

In developing Asia, the world's leader in the production of many vegetable crops, such species face competition from many weeds. Developing Asia leads the world in the production of tomatoes (*Lycopersicon esculentum* Miller) with some of its cultivars being sensitive to increased UV-B, CO_2 or O_3. Only *Cyperus rotundus* L. (a weed) might be generally expected to grow better in competition with vegetable crops, because of increased UV-B. On the other hand, *Chenopodium album* could increase due to elevated CO_2 concentrations and *Portulaca oleracea* subsp. *sativa* (L.) Schuebler & Martens (a weed) could expand into new geographic areas and increase in abundance as a troublesome weed, if climate warming occurs.[19]

Another world-leading crop in developing Asia, tobacco (*Nicotiana tabacum* L.) could find more competition from the perennial monocot weed *Cynodon dactylon*. Additionally, with an increase in air temperature, *C. dactylon* could extend its geographic distribution into new areas that have so far excluded it.[19]

Rice (*Oryza sativa* L.) cultivation in developing Asia, a leading producer in the world, could be affected because it is one of the top nine crops in sensitivity to changes in UV-B, CO_2, or ozone,[7] while its principal weeds are monocots or sedges. However, the balance of competition is not expected to be seriously influenced by increased UV-B. Nevertheless, earlier and increased germination has been projected for one of the grass weeds, *Echinochloa crus-galli*, if climate warms in the future.[19]

Developing Asia also leads the world in wheat cultivation that might be expected to find decreased competition from the dicot weeds *Polygonum convolvulus* L. and *Stellaria media* under increased plant effective UV-B. But, the latter weed species has been projected to increase in its number of life cycles per year, and the consequent seed production, from a longer growing season if the climate warms.[19] Although not statistically significant, except at the 13% level, the weed *Avena fatua* L. (Table 5.2) showed a 3.5% decrease in shoot dry weight, when grown for 34 days as a pure stand in a glasshouse under an estimated plant effective UV-B irradiance of 2.67 W m^{-2}.[14] This was judged to be equivalent to a 20% depletion in the ozone column on 1 June for clear skies at Logan, Utah. In further studies on the competition between wheat and *A. fatua* and between wheat and

Aegilops cylindrica L. grown as mixtures under elevated UV-B, the competitive advantage was found to be with wheat.[21] Changes in the balance of competitive advantage can be very subtle. While it is possible that biomass yield measurements may not show differences, plant morphological changes may. In a field experiment under enhanced UV-B radiation, with a mixture of wheat and *A. fatua*, no change was observed in total biomass, although changes in foliage height growth distribution of wheat over *A. fatua* gave wheat a competitive advantage.[22] Recently, the first well documented mathematical computer model of plant response to UV-B irradiance, confirmed the plant response for the mixtures of the same two species.[23] Unfortunately, the model structure, as well as its validation, is limited to photosynthesis and does not include biomass or harvestable yield.

Likewise, the evaluation given earlier for grape vineyards in developing Asia and transition market countries pertains to conditions from other grape growing countries. Any substantial increase in the UV-B climate might tend not to favor two of its weed species, but any climate warming could by contrast, favor the same two weed species.[19] Results are not available to sort this issue at a quantitative level. The two weed species that cause problems in vineyards are also known to cause problems with cotton cultivation.

Fruit orchards that dominate the same part of the world as grape and cotton, might find less competition from the perennial monocot weed *Agropyron repens* (*Elymus repens* (L.) Gould), if plant-effective UV-B increased during the growing season. But, the herbaceous dicot weed *Portulaca oleracea* could rise in abundance as a troublesome weed, going from one generation to two per year and expand into new geographic areas, if climatic warming occurs.[19]

Developing Africa leads the world in the amount of pasture range. For livestock pastures and meadows, if ambient plant effective UV-B levels increase, there is the possibility that the competitive balance might be tipped against the two herbaceous dicot weeds *Rumex crispus* and *Capsella bursa-pastoris*. But, the latter is also expected to increase, if climate warming occurs.[19]

North America leads the world in corn (*Zea mays* L.) production. Many of the major weeds for field corn are monocots. There is no basis for projecting any change in competition between corn and its weeds under increased exposure to plant effective UV-B irradiance. But, earlier and increased germination has been projected for the weeds *Digitaria sanguinalis* Haller and *Echinochloa crus-galli*. If air temperatures increase, *Cynodon dactylon* might become more widespread.[19]

The limited information for weed competition with citrus and olive crops, both woody dicots, do not raise a concern with regard to any future UV-B exposures. Latin America and Western Europe are the respective world-leading regions of the two species. Here, *Stellaria media* in citrus groves, and *Portulaca oleracea* in competition with olive trees, have both been projected to increase their number of life cycles per year, and consequent seed production from a longer growing season.[19]

With regard to competitive balance only between a crop and its principal weed species, and based on very general patterns that require much more research, if only increased plant effective UV-B levels occur in the future (Table 5.3), the winners will be: sugarcane, some grain crops, vineyards and fruit orchards. Possible increased competition from weeds under higher UV-B levels might make cotton a loser.

But, the outlook might change if one considers possible direct and indirect heating effects from increased CO_2 in the atmosphere on the competitive balance between crops and weeds. In that case, the crop losers might be field corn in North America, citrus orchards in Latin America and the Caribbean, sugar beet and olive orchards in Western Europe, potatoes in Eastern Europe transition market countries, pastures in developing Africa and vegetables, rice, grape vineyards, cotton, fruit orchards and tobacco in developing Asia.

These possible conditions, together with the lack of experimental evidence for the crop and weed combinations shown in Tables 5.1, 5.2 and 5.3 for various parts of the world suggest the need for future research.

References

1. Hale B. UV impacts on food production. In: Heidorn KC, Torrie, B, eds. Proc., International Conf. Ozone Depletion and Ultraviolet Radiation: Preparing for the Impacts, 27-29 April 1994. Victoria, BC, Canada: The Skies Above Foundation, 1995:171-177.
2. Altieri MA. Weed ecology. In: Agroecology—The Scientific Basis of Alternative Agriculture. Boulder, CO: Westview Press, 1984:173-185.
3. Hilbig W, Bachthaler G. Changes in the segetal vegetation dependent on the land-use systems in Germany from 1950 to 1990. 1. Development of photographic techniques, disappearance of weeds, decrease in indicator plants for lime, acidity and moisture and in on-

ion and bulb geophytes—reduction in number of species. Angew Bot 1992; 66(5-6):192-200.
4. Hilbig W, Bachthaler G. Changes in the segetal vegetation dependent on the land-use systems in Germany from 1950 to 1990. 2. Increase in herbicide tolerant species, nitrophilic species and weeds—Increase in presence of undesired rhizome and root—Occurrence and spread of neophytes—Promotion of endangered weeds on arable land—Integrated cultivation. Angew Bot 1992; 66(5-6):201-209.
5. Runeckles VC, Krupa SV. The impact of UV-B and ozone on terrestrial vegetation. Environ Pollut 1994; 83:191-213.
6. Ballare CL, Scopel AL, Sanchez RA et al. Photomorphogenic processes in the agricultural environment. Photochem Photobiol 1992; 56(5):777-788.
7. Krupa SV, Kickert RN. The Greenhouse Effect: Impacts of ultraviolet-B (UV-B) radiation, carbon dioxide (CO_2), and ozone (O_3) on vegetation. Environ Pollut 1989; 61(4):263-393.
8. Krupa SV, Kickert RN. The Effects of Elevated Ultraviolet (UV)-B Radiation on Agricultural Production. A peer reviewed critical assessment report submitted to the Formal Commission on "Protecting the Earth's Atmosphere" of the German Parliament, Bonn, Federal Republic of Germany, 1993.
9. Day TA, Vogelmann TC, De Lucia EH. Are some plant life forms more effective than others in screening out ultraviolet-B radiation? Oecologia 1992; 92(4):513-519.
10. van de Staaij J, Rozema J, Stroetenga M. Expected changes in Dutch coastal vegetation resulting from enhanced levels of solar UV-B. In: Beukema JJ, Wolff WJ, Brouns JJWM, eds. Expected Effects of Climatic Change on Marine Coastal Ecosystems. Netherlands: Kluwer Academic Publ., 1990:211-217.
11. Krupa SV, Kickert RN. The Greenhouse Effect: The impacts of carbon dioxide (CO_2), ultraviolet-B (UV-B) radiation and ozone (O_3) on vegetation (crops). Vegetatio 1993; 104/105:223-238.
12. Cann JC. A comparison of the ecophysiological responses of *Amaranthus retroflexus* and *Chenopodium album* to the exclusion of ultraviolet-A and UV-B radiation in the field and the glasshouse. M.S. thesis. Laramie: University of Wyoming, 1995.
13. Musil CF Differential effects of elevated ultraviolet-B radiation on the photochemical and reproductive performances of dicotyledonous and monocotyledonous arid-environment ephemerals. Plant Cell Environ 1995; 18:844-854.
14. Barnes PW, Flint SD, Caldwell MM. Morphological responses of crop and weed species of different growth forms to ultraviolet-B radiation. Amer J Bot 1990; 77(10):1354-1360.
15. Barnes PW, Ballare CL, Caldwell MM. Photomorphogenic effects of UV-B radiation on plants: Consequences for light competition. J Plant Physiol 1996; 148:15-20.

16. Holm LG, Plucknett DL, Pancho JV et al. The World's Worst Weeds. Honolulu: University Press of Hawaii, 1977.
17. Holm L, Doll J, Holm E et al. World Weeds: Natural Histories and Distribution. New York: John Wiley & Sons, 1997.
18. Hanf M. Ackerunkraeuter Europas. Munich: BLV Verlag, 1990.
19. Ketner P. Will there be a weed problem as a result of climate change? In: Goudriaan J, van Keulen H, van Laar HH, eds. The Greenhouse Effect and Primary Productivity in European Agro-Ecosystems. Wageningen, The Netherlands: Pudoc, 1990:18-19.
20. Fox FM, Caldwell MM. Competitive interaction in plant populations exposed to supplementary ultraviolet-B radiation. Oecologia 1978; 36:173-190.
21. Gold WG, Caldwell MM. The effects of ultraviolet-B radiation on plant competition in terrestrial ecosystems. Physiol Plant 1983; 58:435-444.
22. Barnes PW, Jordan PW, Gold WG et al. Competition, morphology and canopy structure in wheat (*Triticum aestivum* L.) and wild oat (*Avena fatua* L.) exposed to enhanced ultraviolet-B radiation. Funct Ecol 1988; 2:319-330.
23. Ryel RJ, Barnes PW, Beyschlag W et al. Plant competition for light analyzed with a multispecies canopy model. I. Model development and influence of enhanced UV-B conditions on photosynthesis in mixed wheat and wild oat canopies. Oecologia 1990; 82:304-310.

CHAPTER 6

Pathogen and Pest Incidence on Crops

Introduction

Abiotic and biotic factors constrain crop production worldwide.[1] In principle, integrated crop management practices emphasize a reduction in the effect of these constraints with the goal of achieving sufficiency in food production. Independent of this, given the present day socio-political and economic considerations and the variability in the environment, there is a great deal of geographic patchiness in the sufficiency of crop production.[2]

Crop growth and productivity should be viewed as a product of the crop interaction with the total environment. Such an environment consists of both biotic and abiotic factors of the atmosphere and the lithosphere. Traditionally, the study of the effects of abiotic factors on crops has been the domain of agronomists (in some cases, horticulturists), plant physiologists and soil scientists. Similarly, the study of disease-causing pathogens and insect pests has been the domain of phytopathologists and entomologists. Ecologists have provided a degree of overlap among these biological specialties. Only in recent years, however, have these multiple disciplines been related to atmospheric sciences (e.g., air pollution and global climate change). This paucity in disciplinary integration over time has resulted in a significant lack of understanding of the effects of elevated UV-B levels or one of its immediate atmospheric regulators—ozone—on crop pathogen and insect pest incidence.[3]

Stress-crop response relationships are inherently stochastic in nature. Therefore, the spatial and temporal dynamics of various growth regulating variables need to be understood and quantified, before their joint effects on crop growth and productivity can be

---

Elevated Ultraviolet (UV)-B Radiation and Agriculture, by Sagar V. Krupa, Ronald N. Kickert and Hans-Jürgen Jäger.
© 1998 Springer-Verlag and Landes Bioscience.

assessed with confidence. UV-B research has not reached this level at the present time.[4,5]

Secondly, in general, methodology for the exposure of crops to elevated UV-B under field conditions has not been completely satisfactory (chapter 2). Therefore, our present meager knowledge of the effects of elevated UV-B levels on plant pathogens and insect herbivores is based on a few, fragmentary, short-term studies. The results from some of these studies are summarized in the following discussion. The reader is also referred to chapter 7 for supplementary information. In addition, because of some similarities in their mechanism of action (chapter 4), the effects of both elevated UV-B and ozone on crop pathogens and pests are reviewed here.

Elevated UV-B Levels and Incidence of Pathogens

According to Runeckles and Krupa,[5] available evidence clearly shows that the effects of UV-B on the incidence and development of pathogen-induced diseases on crop plants is dependent upon the crop cultivar and age, pathogen type (obligate versus facultative biotrophs), pathogen inoculum level, type of plant organ infected, and the timing and duration of the elevated UV-B exposure.

Orth et al[6] exposed three cultivars of cucumber (*Cucumis sativus* L.) to a daily dose of 11.6 kJ m^{-2} (equal to 16% reduction in stratospheric ozone at the study location, 39°N) biologically effective UV-B radiation in an unshaded greenhouse before and/or after inoculation with *Colletotrichum laginarium* (anthracnose) or *Cladosporium cucumerinum* (scab). Pre-inoculation exposure of 1 to 7 days to UV-B resulted in greater disease severity from both pathogens on the susceptible cultivar, Straight-8. Post-inoculation UV-B exposure was much less effective. Although the resistant cultivars Poinsette and Calypso showed increased severity of anthracnose under a heavy load of pathogen inoculum, when exposed to UV-B (both pre- and post-inoculation), this effect was observed only on cotyledons and not on leaves, where the disease resistance mechanism(s) might be more fully operative (Fig. 6.1).

From their UV-B exposure and crop cultivar studies, Biggs et al[7] suggested that rust disease on wheat (*Triticum aestivum* L.) is more likely to exhibit increased severity when a susceptible, rather than a resistant cultivar is exposed to UV-B radiation. In contrast, severity of Cercospora leaf spot on clonally propagated sugar beet (*Beta vulgaris* L.) increased with exposures to elevated UV-B.[8]

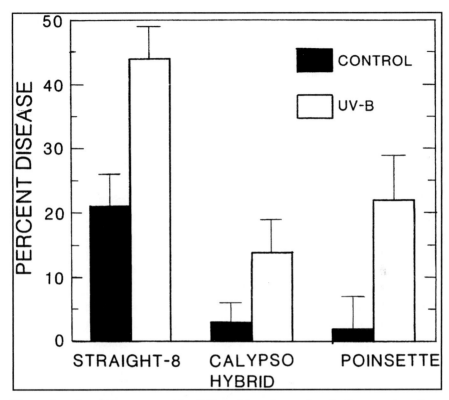

Fig. 6.1. Severity of disease caused by *Colletotrichum lagenarium* (anthracnose) on three cucumber (*Cucumis sativus* L.) cultivars following treatment for 7 days with UV-B (11.6 effective kJ m^{-2}). Plants were inoculated with 1.0 x 10^5 spores ml^{-1} at the end of the UV-B treatment and rated 4 days after inoculation (18 days after planting). Increase in disease severity was restricted to cotyledons on Calypso Hybrid and Poinsette (resistant cultivars), but occurred on both cotyledons and true leaves of Straight-8 (susceptible cultivar). The main effect of UV treatment was significant at p ≤ 0.0001. Reprinted with permission from: Orth AB, Teramura AH, Sisler, HD. Amer J Bot 1990; 77:1188-1192. © 1990 American Journal of Botany.

Owens and Krizek[9] showed that *Cladosporium cucumerinum* spore germination was significantly inhibited by UV-B radiation. Similarly, conidial germ tubes of *Diplocarpon rosae* appear to be sensitive to UV-B prior to the penetration of rose (*Rosa* spp.) leaves.[10] These types of direct adverse effects of UV-B on microorganisms are well known.[11,12]

Sporulation and spore viability in the rice blast fungus (*Magnaporthe grisea*) has been shown to decrease significantly after short-term exposures to realistic levels of UV-B radiation.[13]

Ultraviolet-B was also shown to increase mutation in a pigment gene at nonlethal levels. This could be more significant as most of the fungus is not expected to be exposed directly to UV-B in the host and a mutation in a small portion of the population could result in the spread of that population throughout the crop. However, the non-mutation effects were incorporated by Luo et al[14] in a computer simulation model (RICE-BLAST) to mimic the effects of increased UV-B and air temperature on the rice blast fungus, the spread of the disease and their effect on crop yields in the Philippines. Yield loss from increased UV-B alone, without blast disease or temperature change, was simulated to be about 9 to 10%. But, yield loss caused by the blast, together with UV-B was generally at 15 to 20% under most temperature changes, except at -3°C. To our knowledge, this is the first attempt to explicitly incorporate the effects of increased UV-B radiation on changes in biomass in a plant community, through the use of a computer simulation model.

As to the effects of elevated UV-B on the host versus the pathogen, Carns et al[15] observed that when the anthracnose-resistant cucumber cultivar Poinsette was exposed to UV-B doses injurious to the crop, the mycelial growth of *Colletotrichum laginarium* was partially inhibited and spore germination was severely decreased. Orth et al[6] in their studies on cucumber and *Colletotrichum* concluded that UV-B action on the host was apparently more important than on the fungus per se, since there was no difference in the disease severity between plants that received only pre-inoculation UV-B treatment and those that received both pre- and post-inoculation treatment. However, Ayres et al[16] considered that experimental design to be unsatisfactory, because the control cases had no UV-B exposure and the results do not bear directly upon natural conditions.

In contrast to field-grown plants, a rather artificial laboratory experiment was performed on the effects of extracts of rough lemon (*Citrus jambhiri* Lush.) foliage collected from plants grown for 95 days under 10.2 kJ (BE_{300}) UV-B, on two fungal root rot (*Fusarium solani* and *F. oxysporum*) and two fruit rot (*Penicillum italicum* and *P. digitatum*) pathogens.[17] Under these conditions the phototoxicity of the leaf extracts to the fungi was reduced, as were the furanocoumarin levels in the leaf tissue. The chemo-physiology of a host-fungus system is complex, and much more information is needed before such results can be applied at the agro-ecological level.

The effect of increased UV-B on the damping-off fungus, *Fusarium oxysporum* in spinach (*Spinacia oleracea* L.) was studied in a growth chamber by Naito et al.[18] Over 15 days, with UV-B irradiance of 1.0 W m^{-2} and at two different visible light levels, it was found that the incidence of disease increased up to 70 to 80% compared to about 40% for plants not grown under UV-B. Typical growth reductions for above-ground plant organs were found under the UV-B treatments.

The effects of UV-B on powdery mildew (*Uncinula necator*) on excised leaf disks of grape (*Vitis vinifera* L.), were studied by Willocquet et al.[19] Under ambient sunlight exposure and with and without exclusion of UV-B, it was found that spore germination and mycelial growth were inhibited by sunlight with UV-B compared to the treatment without UV-B. It is not possible to determine if the effect was directly on the fungus, on the host plant, or both. More realistic ecological experiments on intact vines under natural and enhanced sunlight remain to be performed.

An excellent state-of-the-science review of the existing evidence and theory-based effects of increased carbon dioxide, tropospheric ozone and ground-level UV-B radiation on plant disease is found in Manning and von Tiedemann.[3] Table 6.1 lists in general terms the major changes in plant host structure and function and the likely results of increased UV-B on disease incidence and epidemiology. The chain of events triggered by either increased CO_2, increased UV-B or O_3 are shown in Figure 6.2. The current state of knowledge of the effects of increased UV-B on biotic diseases of crop plants is given in Table 6.2, as summarized from Manning and von Tiedemann.[3]

Elevated Ozone Levels and Incidence of Pathogens

A range of response is true for the interactions of plant pathogens and O_3.[20,21] Effects on the host plant, on the pathogen, or on both, might lead to stimulation or inhibition of disease incidence or severity.[22] Dowding[23] has stressed the critical importance of the co-incidence of the timing of pollutant exposure and the pathogen infective period to any effect of O_3 on the establishment of disease.

As with UV-B, in the case of many fungal pathogens, potential effects of exposure to O_3 at the spore stage appear to be minimal, but the organisms are vulnerable following deposition, since their carbohydrate energy reserves are rapidly depleted upon spore germination. The diverse interactions with O_3 on the leaf surface have been

Table 6.1. Potential impacts of major changes in plant structure and function caused by increased UV-B radiation on biotic plant diseases[3]

Major plant responses to increased UV-B	UV-B effects on disease incidence and epidemiology
Structural • Stunted height growth • Increased branching • Increased leaf size	• Microclimate conditions improved for bacterial or fungal infections, and build-up of epidemics
Functional • Reduced net photosynthesis • Increased production of photoactive phenolics (flavonoids) • Premature ripening and senescence • Increased soluble protein content • Decreased membrane lipid content • Sensitivity is relatively high • Adaptability is medium	• Reduction of "high sugar diseases" (rusts, mildews) • Potential antibiotic protection against pathogens • Shortened biotroph infection period, prolonged necrotroph infection period • Improved host nutrient availability, increased growth of necrotrophic pathogens; significant effects on sporulation • Sensitivity is relatively low • Adaptability is generally high
Major plant responses to increased O_3 • Considerable effects on photosynthesis, carbon allocation, secondary metabolism and yield	O_3 effects on disease incidence and epidemiology • Significant effects only at concentrations > 100-250 ppb • Coincidence with sensitive stages is low
Major plant responses to increased CO_2 • Considerable increase of growth and biomass by C3 plants	CO_2 effects on disease incidence and epidemiology • Significant effects mostly at > 0.02-0.05% CO_2 above ambient

discussed by Dowding[23] and include effects on cuticular chemistry, surface properties and exuded materials and on stomatal responses. The growth and development of the pathogen on the host may be inhibited directly by O_3, since toxicity has been observed in axenic culture.[24] Alternatively, important O_3-induced changes in the host may have profound effects on the successful growth and development of the pathogen. Similar interactions may be involved with

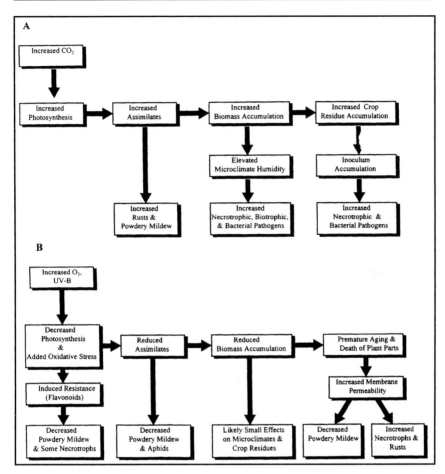

Fig. 6.2. (A) Plant growth and disease susceptibility enhancement from increased ambient CO_2; (B) plant growth and disease susceptibility changes possible from increased ambient tropospheric ozone and/or increased ground-level UV-B radiation.[3]

bacterial pathogens, although infection is usually dependent upon successful entry of the bacteria into the host tissue via wounds or insect vectors.

The development of the pathogen may affect the susceptibility of the host plant to O_3. Since the early report of protection against smog injury to bean (*Phaseolus vulgaris* L.) or sunflower (*Helianthus annuus* L.) leaves infected with the rust fungi, *Uromyces* (*phaseoli*) *appendiculatus* and *Puccinia helianthi*, respectively,[25] there have been numerous reports of infection with viruses, bacteria and fungi, leading to reduced host susceptibility to O_3. Conversely, enhancement of

Table 6.2. Effects of UV-B radiation on diseases of crop plants

Crop	Increased pathogen	Decreased pathogen
Cucumber	Botrytis cinerea, Cladosporium cucumerinum, Colletotrichum lagenarium, Sclerotinia sclerotiorum	Colletotrichum lagenarium
Tomato	Botrytis cinerea	—
Eggplant	Sclerotinia sclerotiorum	—
Greenhouse vegetables	Alternaria dauci, A. porri, A. solani, Botrytis squamosa	—
Oats	—	Puccinia coronata
Beans	—	Uromyces phaseoli
Wheat	Puccinia recondita	Erysiphe graminis
Sugar beets	Cercospora beticola	—
Roses	—	Diplocarpon rosae
Rice	Magnaporthe grisea	—
Squash	Fusarium oxysporum	—
Grape	—	Uncinula necator

the impact of O_3 on the growth of the host plant has been observed with nematode infection.[26] While there is considerable evidence indicating that exposure to O_3 can reduce infection, invasion and sporulation of fungal pathogens, including obligate pathogens such as the rust fungi,[27] examples also exist of increased infection of O_3-injured plants.[28] We do not yet have a clear understanding of the mechanisms involved in O_3-host-pathogen-environment interactions. However, one generalization that can be made is that pathogens which can benefit from injured host cells and disordered transport mechanisms will be enhanced by earlier exposure of the host to O_3, while those that depend on healthy host tissue will be at a disadvantage.[23]

Elevated Ozone or UV-B Levels and Insect Pests

Herbivorous insects and spider mites are major causes of crop loss, but little is known of the effects of UV-B on plant-insect interactions, although there is a sizable body of information in this context about air pollutant effects.[29] As with pathogens and disease, the topic can be subdivided into the influence of the stress agent (UV-B, O_3) on insect attack and population dynamics (whether direct or mediated by changes induced in the plant), and the converse effects of insect attack on plant response to the abiotic stress.

For the record, during the 1970s, experimental laboratory studies were carried out on the responses of a few agricultural insects to artificially increased UV-B exposures.[30] However, lack of adequate quality control and sufficient amount of relevant data, preclude the derivation of any knowledge from that effort. Most recently in field UV-B exclusion experiments at the University of Buenos Aires, Argentina, with *Datura ferox* L., an annual dicot weed found in soybean (*Glycine max* (L.) Merr.) fields, decreased UV-B exposures below the ambient control were found to increase the host attractiveness to leaf-feeding beetles (Coleoptera).[31] On a 24 hour basis, percent of leaves attacked by the beetles decreased by approximately 50% with UV-B exposures versus without. On the basis of percentage of plants attacked over the summer, from complete UV-B exclusion up to the ambient unfiltered levels, the attacks decreased by a factor of 4 (80% reduction). It must be kept in mind that the quantitative nature of this response might not be linear above current ambient UV-B levels, and the experiments do not provide information as to what might happen with future increases in UV-B radiation.

Another plant host-invertebrate interaction under enhanced UV-B regimes is reported by McCloud and Berenbaum[32] and McCloud.[33] *Citrus jambhiri* Lush. (rough lemon), used as root stock in commercial citriculture (contains furanocourmarins that are highly toxic to insect herbivores), was used in greenhouse UV-B experiments with the cabbage looper (*Trichoplusia ni*) and a daily UV-B dose of 6.4 kJ, BE_{300}. Under the enhanced UV-B exposure, furanocourmarins increased in concentration in the plants, larvae as caterpillars developed more slowly over time and their survivorship was reduced compared to unenhanced UV-B doses. Thus, elevated UV-B levels tended to increase the protection of the plant against the insect. Similar results were found for the cabbage looper feeding on the weedy forb, *Plantago lanceolata* L. However, no effects of UV-B exposures were found on the host plant for the performance of a specialist invertebrate herbivore, *Junonia coenia*.[33]

Host plant resistance to insect attack may be modified through metabolic changes which affect feeding preference and insect behavior, development and fecundity. Ozone- induced changes in both major and secondary metabolites may be qualitative and quantitative in their nature, while there is abundant evidence that such changes can influence insect growth and development, there have been few investigations of the specific effects of O_3. Trumble et al[34] reported that the tomato pinworm (*Keiferia lycopersicella*) developed faster on O_3-injured tomato (*Lycopersicon esculentum* Mill.) plants, although fecundity and female longevity were unaffected. The Mexican bean beetle (*Epilachna varivestis*) was found to show an increased preference for O_3-treated soybean foliage with increased O_3 exposures.[35] Such preferences can lead to increased larval growth rates, as shown by the work of Chappelka et al.[36] Other examples are reviewed in Runeckles and Chevone.[21]

There is only one report concerning insect attack modifying the effects of O_3 on a herbaceous host. Rosen and Runeckles[37] showed that the combination of extremely low levels of O_3 (0.02 ppm) and infestation with the greenhouse whitefly (*Trialeurodes vaporariorum*) acted synergistically in inducing accelerated chlorosis and senescence of bean leaves. They speculated that the effect might be the result of the reaction of O_3 with enhanced ethylene production resulting from whitefly injury.

Plant response to stress through the accumulation of secondary metabolites (pigments) may be a generalized reaction, although the rate of such accumulation and the biosynthetic pathway leading to the production of the accumulated product may vary with the type of stress.[38] Acute or episodal exposures to ozone (relatively high concentrations from a few consecutive hours to days) can result in localized pigment accumulation in the foliar tissue, and subsequently, premature senescence and leaf abscission.[39] In comparison, the modulated response of crops to increasing levels of UV-B can result in generalized pigment accumulation in the epidermal cell layer and this can offer protection against additional UV-B.[40] In both cases, one can speculate that the pigment accumulation may alter insect herbivory.[41] Nevertheless, depending on the sequence, timing and extent of various forms of stress factors, crop response will be highly variable in time. Thus, in summary the information available on the biotic interactions involving UV-B or O_3 is fragmentary and precludes a clear unraveling of the complexities of the relationships. This is a situation that can only be remedied by further systematic investigation.

References
1. Oerke E-C, Dehne H-W, Schönbeck F et al. Crop Production and Crop Protection: Estimated Losses in Major Food and Cash Crops. Amsterdam: Elsevier, 1994.
2. FAO. http://apps.fao.org/lim500/nph-wrap.pl?Production.Crops. Primary&Domain=SUA, 1997.
3. Manning WJ, von Tiedemann A. Climate change: Potential effects of increased atmospheric carbon dioxide (CO_2), ozone (O_3), and ultraviolet-B (UV-B) radiation on plant diseases. Environ Pollut 1995; 88:219-245.
4. Krupa SV, Kickert RN. The Greenhouse Effect: The impacts of carbon dioxide (CO_2), ultraviolet-B (UV-B) radiation and ozone (O_3) on vegetation (crops). Vegetatio 1993; 104/105:223-238.
5. Runeckles VC, Krupa SV. The impact of UV-B radiation and ozone on terrestrial vegetation. Environ Pollut 1994; 83:191-213.
6. Orth AB, Teramura AH, Sisler HD. Effects of ultraviolet-B radiation on fungal disease development in *Cucumis sativus*. Amer J Bot 1990; 77:1188-1192.
7. Biggs RH, Webb PG, Garrard LA et al. The Effects of Enhanced Ultraviolet-B Radiation on Rice, Wheat, Corn, Citrus and Duckweed. Environmental Protection Agency Interim Report 80: 8075-03. Washington, DC: U.S. Environmental Protection Agency, 1984.
8. Panagopoulos I, Bornman JF, Björn LO. Response of sugar beet plants to ultraviolet-B (280-320 nm) radiation and Cercospora leaf spot disease. Physiol Plant 1992; 84:140-145.
9. Owens OVH, Krizek DT. Multiple effects of UV radiation (265-330 nm) on fungal spore emergence. Photochem Photobiol 1980; 32:41-49.
10. Semeniuk P, Stewart RN. Effect of ultraviolet (UV-B) irradiation on infection of roses by *Diplocarpon rosae* Wolf. Environ Exp Bot 1981; 21:45-50.
11. Sussman AS, Halvorsen HO. Spores. New York: Harper & Row, 1966.
12. Leach CM. A practical guide to the effects of visible and ultraviolet light on fungi. Methods Microbiol 1971; 4:609-664.
13. Leung H, Christian D, Loomis P et al. Effects of ultraviolet-B irradiation on spore viability, sporulation, and mutation of the rice blast fungus. In: Peng S, Ingram KT, Neue H-U et al, eds. Climate Change and Rice. New York: Springer, 1995:158-168.
14. Luo Y, TeBeest DO, Teng PS et al. Risk analysis of rice leaf blast epidemics associated with effects of enhanced ultraviolet-B and temperature changes in the Philippines. IRRN 1994; 19:57-58.
15. Carns HR, Graham JH, Ravitz SJ. Effects of UV-B radiation on selected leaf pathogenic fungi and on disease severity. In: UV-B Biological and Climatic Effects Research (BACER), FY 77-78 Research Report on Impacts of Ultraviolet-B Radiation on Biological Systems: A Study Related to Stratospheric Ozone Depletion. Final Report, Vol. I, EPA-IAG-6-0168. Washington, DC: USDA EPA, Stratospheric Im-

pact Research and Assessment Program (SIRA), U.S. Environmental Protection Agency, 1978.
16. Ayres PG, Gunasekera TS, Rasanayagam MS et al. Effects of UV-B radiation (280-320 nm) on foliar saprotrophs and pathogens. In: Frankland JC, Magan N, Gadd GM, eds. Fungi and Environmental Change. New York: Cambridge University Press, 1996:32-50.
17. Asthana A, McCloud ES, Berenbaum MR et al. Phototoxicity of *Citrus jambhiri* to fungi under enhanced UV-B radiation: Role of furanocoumarins. J Chem Ecol 1993; 19:2813-2830.
18. Naito Y, Honda Y, Kumagai T. Effects of supplementary UV-B radiation on development of damping-off in spinach caused by the soil-borne fungus *Fusarium oxysporum*. Mycoscience 1996; 37:15-19.
19. Willocquet L, Colombet D, Rougier M et al. Effects of radiation, especially ultraviolet-B, on conidial germination and mycelial growth of powdery mildew. Eur J Plant Path 1996; 102:441-449.
20. Chappelka AH, Chevone B. Tree responses to ozone. In: Lefohn AS, ed. Surface Level Ozone Exposures and Their Effects on Vegetation. Chelsea, MI: Lewis Publishers, 1992:271-324.
21. Runeckles VC, Chevone B. Crop responses to ozone. In: Lefohn AS, ed. Surface Level Ozone Exposures and Their Effects on Vegetation. Chelsea, MI: Lewis Publishers Inc., 1992.
22. Heagle AS. Interactions between air pollutants and parasitic plant diseases. In: Unsworth MH, Ormrod DP, eds. Effects of Gaseous Air Pollution on Agriculture and Horticulture. London: Butterworth Scientific, 1982:333-348.
23. Dowding P. Air pollutant effects on plant pathogens. In: Schulte-Hostede S, Darrall NM, Blank LW et al, eds. Air Pollution and Plant Metabolism. London: Elsevier, 1988:329-355.
24. Krause CR, Weidensaul TC. Effects of ozone on the sporulation, germination and pathogenicity of *Botrytis cinerea*. Phytopathology 1978; 68:196-198.
25. Yarwood CE, Middleton JT. Smog injury and rust infection. Plant Physiol 1954; 29:393-395.
26. Bisessar S, Palmer KT. Ozone, antioxidant spray and *Meloidogyne hapla* effects on tobacco. Atmos Environ 1984; 18:1025-1027.
27. Heagle AS. Interactions between air pollutants and plant parasites. Annu Rev Phytopathol 1973; 11:365-388.
28. Manning WJ, Feder WA, Perkins I et al. Ozone injury and infection of potato leaves by *Botrytis cinerea*. Plant Dis Rept 1969; 53:691-693.
29. Manning WJ, Keane KD. Effects of air pollutants on interactions between plants, insects and pathogens. In: Heck WW, Taylor OC, Tingey DT, eds. Assessment of Crop Loss From Air Pollutants. London: Elsevier Applied Science, 1988:365-386.
30. Hayes DK. Influence of broad band UV-B on physiology and behavior of beneficial and harmful insects. In: UV-B Biological and Climatic Effects Research (BACER), FY 77-78 Research Report on Impacts of Ultraviolet-B Radiation on Biological Systems: A Study

Related to Stratospheric Ozone Depletion. Final Report, Vol. I, EPA-IAG-6-0168. Washington, DC: USDA EPA, Stratospheric Impact Research and Assessment Program (SIRA), U.S. Environmental Protection Agency, 1978.
31. Ballaré CL, Scopel AL, Stapleton AE et al. Solar ultraviolet-B radiation affects seedling emergence, DNA integrity, plant morphology, growth rate, and attractiveness to herbivore insects in *Datura ferox*. Plant Physiol 1996; 112:161-170.
32. McCloud ES, Berenbaum MR. Stratospheric ozone depletion and plant-insect interactions: Effects of UV-B radiation on foliage quality of *Citrus jambhiri* for *Trichoplusia ni*. J Chem Ecol 1994; 20:525-539.
33. McCloud ES. Stratospheric ozone depletion and plant-insect interactions: Effects of UV-B radiation on generalist and specialist herbivores on a tropical tree and a temperate forb. Ph.D. thesis. Urbana-Champaign: University of Illinois, 1995.
34. Trumble JT, Hare JD, Musselman RC et al. Ozone-induced changes in host plant suitability: Interactions of *Keiferia lycopersicella* and *Lycopersicon esculentum*. J Chem Ecol 1987; 13:203-218.
35. Endress AG, Post SL. Altered feeding preference of Mexican bean beetle *Epilachna varivestis* for ozonated soybean foliage. Environ Pollut 1985; 39:9-16.
36. Chappelka AH, Kraemer ME, Mebrahtu T et al. Effects of ozone on soybean resistance to the Mexican bean beetle (*Epilachna varivestis* Mulsant). Environ Exp Bot 1988; 28:53-60.
37. Rosen PM, Runeckles VC. Interaction of ozone and greenhouse whitefly in plant injury. Environ Conserv 1976; 3:70-71.
38. Kosuge T, Kimpel JA. Energy use and metabolic regulation in plant-pathogen interactions. In: Ayres PG, ed. Effects of Disease on the Physiology of the Growing Plant. Cambridge, England: Cambridge University Press, 1981:29-45.
39. Krupa SV, Manning WJ. Atmospheric ozone: Formation and effects on vegetation. Environ Pollut 1988; 50:101-137.
40. Caldwell MM, Teramura AH, Tevini M. The changing solar ultraviolet climate and the ecological consequences for higher plants. Trends Ecol Evolut 1989; 4:363-367.
41. Harborne JB, ed. The Flavonoids: Advances in Research Since 1980. London: Chapman & Hall, 1988.

CHAPTER 7

Integrated View of Environment-Crop Interactions

Introduction

Some of the topics addressed in this book relate to the application of current scientific knowledge to the global spatial grid for revealing patterns of spatial variability in the present day world. Other topics are more fundamental for enabling a comprehensive understanding of the dynamic relationships that occur at any spatial level, from global, down to the local scale. The contents of this chapter should be interpreted in the latter context without consideration of the spatial term. The focus is on the combination of factors interacting and impacting the production of a given crop.

The joint effects of elevated UV-B, CO_2, surface O_3, heat, moisture, disease and insect pests on crops is the topic under which all potential limitations to crop production should be integrated. Such an effort requires a dynamic systems perspective, and ideally, it would be implemented as computer-based technological information-decision support systems. However, the state of our current knowledge is still too primitive to implement such an approach with any confidence. Experimental and observational findings of the joint effects of the aforementioned variables and their interactions on plant responses are fragmented and are only beginning to be studied in a quantitative manner. Because of this complexity, the topic of this chapter often involves a discussion of modeling and statistical data analysis more so than the subject matter of most of the other chapters in this book.

Over the years, field-grown crops have been exposed to increases in CO_2, extremes of temperature, moisture, and light, and fluctuations in nutrients—all factors which represent resources

Elevated Ultraviolet (UV)-B Radiation and Agriculture, by Sagar V. Krupa, Ronald N. Kickert and Hans-Jürgen Jäger.
©1998 Springer-Verlag and Landes Bioscience.

needed for growth. In addition, crops have been and are being exposed to increases in tropospheric O_3 and perhaps in the future, increases in ground-level UV-B. Added to these interactions are the effects of plant diseases, insect pests and competition from weeds. Therefore, agricultural production is in part a result of the ability of a crop to grow under the impacts of these multiple stress factors.

Not only with respect to increases in UV-B, but also with regard to other factors of climate change and the consequences for biodiversity, there is considerable interest in how plants respond to multiple environmental stresses. It is recognized that much research has been done on individual factors constraining growth, but in recent years, an increasing number of studies are being aimed at examining possible interactions between the multiple factors.[1]

Potential Combinations of Interactions

From the previous narrative, a minimum of ten potential stressor categories can be identified and a combination of those factors could interact to affect some dependent crop response of interest. It should be recognized that this is a 10-dimensional problem, and is 11 dimensions when a dependent plant response variable is included. The total number of combinations of N parameters taken as some number at a given time is: 2^n-1. For a set of 10 potential stress agents, this means 1,024 potential interactions. Subtracting the 10 single factorial relationships to a dependent response variable, leaves a possible total of 1,012 potential interactions of the combinations of two or more stress agents affecting a single crop response. Clearly, this is a daunting challenge for crop ecologists. It is easy to display the interactions in a graphic form for any 3 of the 11 dimensions (including the dependent response variable) and thus, a larger number of interactions cannot be illustrated.

Types of Interactions

Synergistic change occurs when two or more environmental or ecological processes jointly interact, either simultaneously or sequentially, in a way that the result is not a simple sum of the otherwise individual responses, but instead is multiplicative. This consists of an amplification or dampening of effects and a compounding of the impacts. Sometimes the tolerance of a species to one source of stress becomes reduced when other stresses are being experienced simultaneously.[2] Very little is known about such ecological synergistic interactions and the mechanisms responsible for them.

Biologically, the value of harvestable crop parts for food is a result of stored energy (carbon, C), along with the nutrient content. For the purposes of national and international policy analysis and planning, first considerations are probably given to the production of the bulk amount of edible crop part. An essential consideration is: "What might happen to the crop productivity, namely the amount of carbon stored?" Although there are elaborate models of crop growth processes, when viewed at the most fundamental and general level, the growth of any crop is a balance between carbon uptake, and subsequent carbon storage, remaining after the necessary losses. This is similar to a financial budget: income is disbursed into savings, after losses through expenses. The same line of reasoning can be used for nutrients such as nitrogen, but that aspect will be considered only in a limited way in the following discussion. Plants absorb carbon from atmospheric carbon dioxide during the daytime (photosynthesis), and they normally lose varying portions of that carbon both during the day (photorespiration) and at night (dark respiration):

$$\text{C-Uptake} = (\text{C-Storage}) - (\text{C-Losses}) \qquad (1)$$

In addition to the normal maintenance costs, with plant stress from unexpected increases in environmental contaminants, additional carbon losses will occur within the crop as it attempts to repair the stress effect.

If carbon uptake increases due to increased atmospheric CO_2, and possibly due to climate warming, and if carbon losses also increase, because of episodes of increased UV-B radiation interspersed between episodes of tropospheric ozone, it is conceivable that the net effect on carbon storage might only be a negligible change in the balance:

$$\text{C-Uptake (increase)} = [\text{C-Storage (little change)}] - [\text{C-Losses (increase)}] \qquad (2)$$

But, even with a direct fertilization effect of increased ambient CO_2, if carbon losses increase substantially, because of increases in atmospheric contaminants and perhaps from changes due to less optimal levels of meteorological resources needed for crop growth, then carbon storage in plant parts could decrease, i.e., decreased crop production:

C-Uptake (increase) = [C-Storage (decrease)] - [C-Losses (increase)]   (3)

This idealized concept would vary according to the differences in the sensitivities of different crop species and cultivars to multiple, simultaneous and/or sequential exposures to different stress factors. Within an optimal range for crop growth, light, heat, moisture and nutrients are necessary resources. The soil nutrient status is usually managed for a given soil fertility level by fertilizer usage practices. But, if decreases in seasonal light levels, or changes in heat and/or moisture occur in a given region (e.g., increase in cloud cover), those changes might be considered as environmentally undesirable, coupled with any increases in UV-B radiation, tropospheric ozone, or increases in "biotic factors" such as weeds, diseases or insect pests. The questions that require attention are: "How can increases in the various undesirable parameters *jointly affect* carbon losses as changes in carbon storage in plant parts?" and "What is the nature of the experimental and/or observational evidence?"

In a previous evaluation of the effects of changes in UV-B radiation on natural ecosystems, a statement was made that the only information available for the combined effects of UV-B and other climatic resources or contaminants affecting plant growth, was for the joint effects of moisture stress and UV-B.[3] A similar limitation in our knowledge was identified in the Proceedings of the Workshop of the Scientific Committee on Problems of the Environment (SCOPE).[4] According to that report, it is known that other environmental factors, such as visible light and water stress, can change a UV-B-induced plant response. Such interactions can be exacerbative or ameliorative. In the short time since these evaluations were made, information on other joint interactions from combined stress effects that can occur from climate change and stratospheric ozone depletion has been published (e.g., van de Staaij et al[5]). These various interactions between elevated UV-B, CO_2, O_3, heat, moisture, disease and insect pests form the basis for the following discussion.

There is a difference between potential *environmental interactions* between two or more physico-chemically triggered processes such as plant growth response to increased CO_2 and increased UV-B (if indeed a true interaction is found between these two factors), and the *species interactions* that can exist between two or more types of living organisms. Examples of the latter are discussed by Billick and Case.[6] Trying to consider the interactions between changes in

Table 7.1. Interactive plant response to elevated UV-B levels as modified by other growth regulating factors

Growth regulating factors	Response to elevated UV-B levels
Higher light	Decreased negative effect
Optimal water availability	Increased negative effect
Increased CO_2	Increased negative effect
Increased air temperature	Increased negative or positive effect
Optimal nutrient availability	Increased (short-term) negative effect
Increased plant diseases	Responses too variable to categorize
Increased insect pests	Unknown response status of infested plant or insect sensitivity
Increased heavy metals	Increased or decreased negative effect

the physico-chemical processes and their resulting effects on plant host and pathogen and/or insect pests, and the interactions between the organismal species themselves in that system, comprises one of the most complex issues because it can involve both types of interactions.

Comprehensive reviews of the responses of vegetation to combined exposures of increased UV-B and different light levels (photosynthetically active radiation, 400-700 nm), moisture availability, atmospheric carbon dioxide, temperature, mineral nutrient availability, diseases, insects, or heavy metals are found in Bornman and Teramura,[7] and Caldwell et al.[8] Some of the case studies, for combinations of UV-B with another environmental factor, were based on experiments with plant species that are botanically far removed (e.g., spruce versus a crop plant). One should be extremely cautious in assuming that results observed for a non-crop plant will apply to a crop species. However, current evidence suggests the general features presented in Table 7.1.

General Methods of Interaction Analysis

The search for interactions, especially in a quantitative way, between two or more potential stress agents, as observed in various plant responses, can be the example of a scientific investigation, and it is therefore, instructive to consider how scientists outside the field of UV-B research have looked for such interactive effects.

One of the best descriptions of such approaches can be found in Kreeb and Chen.[9] The authors describe how to use various combinations of statistical regression analyses with data for two factors and a dependent variate. Using various mathematical combinations of linear, exponential, natural log and power expressions, both linear and nonlinear relationships could be examined. Nine statistical formulations represented algebraically additive relationships in which no interaction term was present. Four multiplicative formulations were discussed wherein interactions were represented. For various sets of data, a test of significance revealed which equation best fitted the experimental data and thereby revealed whether or not there appeared to have been an interaction between the two environmental factors in association with the dependent variate. For example, the response of chlorophyll content of wheat (*Triticum aestivum* L.) leaves to various levels of salt stress and drought was found to fit best with a power-exponential relationship, which indicated that an interaction existed in the form of a dampening effect. The relationship between potential osmotic pressure of cell sap to salt stress and drought also showed an interaction in the form of a synergistic effect. There is much to be learned in this approach with regard to the current interest in searching for interactions between simultaneous UV-B exposures and other environmental factors on crop production variables. Beyond the few mathematical relationships discussed by Kreeb and Chen,[9] it should be recognized that there are desktop computer software (e.g., TableCurve 3D, by Jandel Scientific, San Rafael, CA, U.S.) that enable the exploration and representation of response surfaces for data on three factors from among 243 polynomial equations, 260 rational equations and 172 non-linear models, fitting up to 36,582 equations, and then sorting the best equations according to goodness of fit.

Experimental Results

UV-B, O_3 and Other Stress Factors

Only recently have studies begun to examine plant growth responses under multiple stresses. An analysis of the scientific literature on plant response to individual exposures to elevated UV-B, CO_2, or O_3 indicated that among all crops, nine species might have cultivars that are relatively sensitive to each of the factors (Table 7.2).[17,18] Five of these crops are important in some developing countries of

Table 7.2. Comparison of the sensitivities of agricultural crops to enhanced CO_2 (mean relative yield increases of CO_2-enriched to control) (after Kimball,[10-12] Cure,[13] Cure and Acock[14]), for CO_2 concentrations of 1200 µL L^{-1} or less (Kimball[10,11]), or 680 ppm (Cure and Acock[14]); to enhanced UV-B radiation; and to ground-level O_3. Species considered to be sensitive to all three factors are indicated by shading.[a]

Crop type	Crop[b]	Enhanced CO_2; mean relative yield increase[c]	Sensitivity to enhanced UV-B	Sensitivity to O_3
Fiber crops	Cotton[1]	3.09	Tolerant	Sensitive
C4 grain crops	Sorghum	2.98	Sensitive	Intermediate
Fiber crops	Cotton[1]	2.59-1.95	—	—
Fruit crops	Eggplant	2.54-1.88	Tolerant	Unknown
Legume seeds	Peas	1.89-1.84	Sensitive	Sensitive
Roots and tubers	Sweet potato	1.83	Unknown	Unknown
Legume seeds	Beans	1.82-1.61	Sensitive	Sens./Intermed.
C3 grain crops	Barley[2]	1.70	Sensitive	Tolerant
Leaf crops	Swiss chards	1.67	Sensitive	Unknown
Roots and tubers	Potato[3]	1.64-1.44	Sens./Toler.	Sensitive
Legume crops	Alfalfa	1.57[d,e]	Tolerant	Sensitive
Legume seeds	Soybean[4]	1.55[f]	Sensitive	Tolerant
C4 grain crops	Corn[5]	1.55	Tolerant	Sensitive
Roots and tubers	Potato[3]	1.51	—	—
C3 grain crops	Oats	1.42	Sensitive	Sensitive
C4 grain crops	Corn[5]	1.40[f]	—	—
C3 grain crops	Wheat[6]	1.37-1.26	Tolerant	Intermediate
Leaf crops	Lettuce	1.35	Sensitive	Sensitive
C3 grain crops	Wheat[6]	1.35	—	—

Table 7.2 continued

Crop type	Crop[b]	Enhanced CO_2; mean relative yield increase[c]	Sensitivity to enhanced UV-B	Sensitivity to O_3
Fruit crops	Cucumber	1.43–1.30	Sensitive	Intermediate
Legume seeds	Soybean[4]	1.29	—	—
C4 grain crops	Corn[5]	1.29	—	—
Roots and tubers	Radish	1.28	Tolerant	Intermediate
Legume seeds	Soybean[4]	1.27–1.20	—	—
C3 grain crops	Barley[2]	1.25	—	—
C3 grain crops	Rice[7]	1.25	Sensitive	Intermediate
Fruit crops	Strawberry	1.22–1.17	Unknown	Tolerant
Fruit crops	Sweet pepper	1.60–1.20	Sens./Toler.	Unknown
Fruit crops	Tomato	1.20–1.17	Sensitive	Sens./Intermed.
C3 grain crops	Rice[7]	1.15	—	—
Leaf crops	Endive	1.15	Unknown	Intermediate
Fruit crops	Muskmelon	1.13	Sensitive	Unknown
Leaf crops	Clover	1.12	Tolerant	Sensitive
Leaf crops	Cabbage	1.05	Tolerant	Intermediate
Flower crops	Nasturtium	1.86	—	—
Flower crops	Cyclamen	1.35	—	—
Flower crops	Rose	1.22	Tolerant	—
Flower crops	Carnation	1.09	—	Intermediate
Flower crops	Chrysanthemum	1.06	Tolerant	Intermediate
Flower crops	Snapdragon	1.03	—	—

[a]Reprinted with permission from: Krupa SV, Kickert RN. Environ Pollut 1989; 61(4):263–393. © 1989 Elsevier Science Ltd. [b]Crops with superscript numbers have more than one ranking. [c]From Kimball,[10,11] and, if shown, the second value is from Kimball.[12] [d]Mean relative yield increase of CO_2-enriched (680 ppm) to control crop (300–350 ppm), after Cure and Acock.[14] [e]Based on biomass accumulation; yield not available. [f]Field-based result from Rogers et al.[15,16]

Asia (see chapter 8). Together with their relative ranking in sensitivity to increases in surface UV-B, CO_2, and O_3, they are: pea (*Pisum sativum* L.) (2), bean (*Phaseolus vulgaris* L.) (3), lettuce (*Lactuca sativa* L.) (6), cucumber (*Cucumis sativus* L.) (7) and tomato (*Lycopersicon esculentum* Miller) (9). Potato (*Solanum tuberosum* L.) is an important crop in the transition market countries of Eastern Europe and this crop ranked 4th in its sensitivity. Rice (*Oryza sativa* L.), ranking 8th in sensitivity to each of the three factors, is important in developing Asian countries. The transition market countries of Eastern Europe lead the world in oat (*Avena sativa* L.) production which ranked 5th in the order of its sensitivity among all the crops examined. The highest ranking crop, sorghum (*Sorghum vulgare* Moench), is important in developing African countries.

An evaluative review of published experimental data has been produced for what is known about the joint effects of surface UV-B, O_3, CO_2, other contaminant gases, light, moisture, plant diseases and insect herbivores.[19] This review concluded that: (1) many growth chamber studies are flawed by using light levels that are far less than the ambient; this mitigates against photorepair processes in the plants and can show a higher adverse effect in the plant response to UV-B exposure than would occur in an ambient environment; (2) soybean (*Glycine max* (L.) Merr.) vegetative growth can show a less-than-additive biomass response to combined exposure to UV-B and O_3;[20] (3) in some annual crop plants, sequential exposure to UV-B followed by O_3 can show a negative (adverse), less-than-additive response in pollen tube growth,[21] important for seed production; (4) under water stress, some plants can become less sensitive to UV-B and become more sensitive to water stress; (5) little is known about any possible interaction between plant nutrient stress and response to UV-B exposure; (6) a 43% increase in ambient CO_2 levels can lessen the growth and physiological effects of O_3 exposure to soybean;[22] (7) in general, plant injury from O_3 tends to increase under nitrogen stress, but plant O_3 susceptibility often decreases under water stress; (8) while there is some knowledge of plant disease response to UV-B, very little is known about actual plant responses to the interaction of biotic pathogens and UV-B (see chapter 6); (9) even less is known about actual plant responses to the interaction of insect herbivores, UV-B, and other physical, chemical and biotic stress factors (see chapter 6); and (10) there have been numerous reports of reduced plant susceptibility to O_3, being conferred by host infection of viruses, bacteria and fungi.[19]

UV-B-CO$_2$ interactions

General plant response patterns show that increased CO$_2$ stimulates plant growth, while increased UV-B results in a decrease (Table 7.3). One study has concluded that the combined effects appear to be additive.[23] Growth rate of wheat increased from increased CO$_2$ alone, but was reduced when growth occurred under a combination of both elevated UV-B and CO$_2$.

An experimental glasshouse study examined pea (*Pisum sativum* L.), tomato (*Lycopersicon esculentum* Mill. cv. Moneymaker), and aster (*Aster tripolium* L.) shoot dry weight response to elevated UV-B radiation and CO$_2$.[24] In that report, the duration of the exposures was not stated and therefore, it is not possible to discern the total exposure dose. The two treatments were 2.8 W m^{-2}, and 0.0 W m^{-2} UV-B, where the first exposure level represented a cloudless midday summer value at Amsterdam, The Netherlands. Hence, the experiment was conducted to determine what happens when UV-B is eliminated, although it is not clear that the statistical approach used was designed to answer that question. If it were to have been an experiment on increased UV-B, then upper levels of 3.5 to 4.0 W m^{-2} should have been used, rather than 2.8. Also, light levels for the growth conditions were relatively low (175 mE m^{-2} sec^{-1}). Since only two treatment levels were used, they represent an insufficient number to determine the quantitative nature of any interaction between CO$_2$ and UV-B on plant response. Two data points can only provide a straight line even if the relationship is not truly linear. At best, four points in three dimensions can only represent a distorted plane. While this would reveal nonlinearity in the response to the interaction, it is far from ideal for identifying any synergistic interactions. A minimum of a 3 x 3 treatment design should have been used, but preferably an even larger one. In spite of this, Figures 7.1A-C show the behavior of the plant response to these limited experimental designs. Both pea and tomato shoot dry weights showed statistically significant changes, decreases from increased UV-B alone and increases from increased CO$_2$ alone. But, when exposed simultaneously to increases in both factors over the same unspecified time period, there were no statistically significant ($p < 0.05$) interactions. The study concluded that the combined effect of UV-B with CO$_2$ was additive, but critical inspection of the results appear to indicate that a positive effect from increased CO$_2$ can be totally eliminated by a predominating negative effect of increased UV-B. A tentative observation might be that

Table 7.3. Overview of the effects of UV-B, CO_2 and O_3 on plants in single-exposure mode[a]

Plant characteristic	Increased UV-B only	Plant response to environmental change Doubling of CO_2 only	Increased tropospheric O_3 only
Photosynthesis	Decreases in many C3 and C4 plants	C3 plants increase up to 100%, but C4 plants show only a small increase	Decreases in many plants
Leaf conductance	Not affected in many plants	Decreases in C3 and C4 plants	Decreases in sensitive species and cultivars
Water use efficiency	Decreases in most plants	Increases in C3 and C4 plants	Decreases in sensitive plants
Leaf area	Decreases in many plants	C3 plants increase more than C4 plants	Decreases in sensitive plants
Specific leaf weight	Increases in many plants	Increases	Increases in sensitive plants
Crop maturation rate	Not affected	Increases	Decreases
Flowering	Inhibits or stimulates flowering in some plants	Earlier flowering	Decreased floral yield, number and yield of fruits, and delayed fruit setting
Dry matter production and yield	Decreases in many plants	C3 plants nearly double, but C4 plants show only small increases	Decreases in many plants
Sensitivity between species	Large variability in response among species	Major differences between C3 and C4 plants	Large variability in sensitivity between species
Sensitivity within species (cultivars)	Response differs between cultivars of a species	Can vary among cultivars	Response differs between cultivars of a species
Drought stress sensitivity	Plants become less sensitive to UV-B, but sensitive to lack of water	Plants become less sensitive to drought	Plants become less sensitive to ozone but sensitive to lack of water
Mineral stress sensitivity	Some plants become less while others more sensitive to UV-B	Plants become less responsive to elevated CO_2	Plants become more susceptible to ozone injury

[a]Reprinted with permission from: Krupa SV, Kickert RN. Environ Pollut 1989; 61(4):263-393. © 1989 Elsevier Science Ltd.

the effect was negative, and less-than-additive, because the CO_2 effect did not appear to compensate for the UV-B induced injury.

It is known from glasshouse experiments with a salt marsh grass (*Elymus athericus* L.), that 4 of 12 growth parameters measured, where response was statistically significant when the plants were grown under either increased CO_2 or UV-B alone, became nonsignificant when the plants were grown simultaneously under increased CO_2 and UV-B.[5] Three other statistically significant growth effects under increased UV-B alone, including a decrease in leaf area, also became nonsignificant when the plants were grown under both increased CO_2 and UV-B. When plants were grown under the influence of increases in both factors, even though changes in all 12 growth parameters were nonsignificant compared to controls (ambient CO_2 levels and near-ambient UV-B), there were indications that 7 of the growth parameters showed antagonistic responses, 2 showed synergistic responses, and 2 showed additive responses from the interactions between simultaneous increase in CO_2 and UV-B. In a similar, but subsequent study,[25] a statistically significant interactive effect with CO_2 and UV-B was observed for total plant dry weight and root dry weight. It is not known whether the growth responses of the particular salt marsh grass would serve as an example of how a rice crop might react under elevated CO_2 and UV-B.

For rice, in a phytotron experiment, an antagonistic interaction was observed between a doubled CO_2 level of the ambient and an increased UV-B radiation regime (approximating a 10% equatorial stratospheric ozone reduction).[26] By itself, a doubled CO_2 concentration significantly elevated the photosynthesis, seed yield and biomass, but these increases disappeared when rice was grown under both elevated levels of CO_2 and UV-B. This finding should be of interest to agricultural planners in the countries of developing Asia.

In the early 1990s, an analysis of plant responses to elevated atmospheric carbon dioxide concluded that there was no observational or experimental evidence of a statistically significant joint interactive effect of increased CO_2 and increased UV-B on plant growth properties, even though there were reasons to expect that such interactions should exist.[27] However, no definition was given of what would comprise such statistically significant evidence. If the amount of change in a plant property under exposure to both independent variables is numerically greater than the algebraically additive changes in that property to each of the variables alone, then

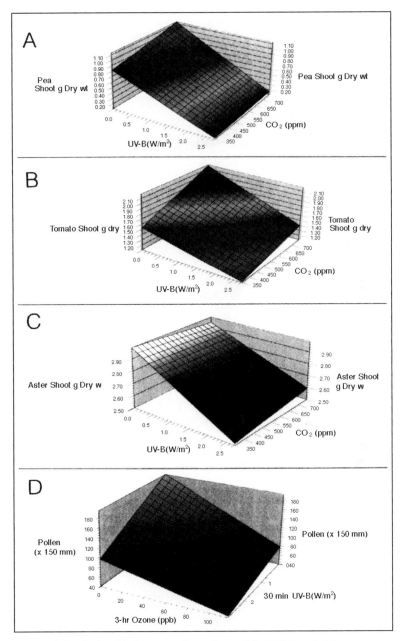

Fig. 7.1. (A) Relationship between shoot dry weight, UV-B and CO_2 exposures in pea (*Pisum sativum* L.).[24] (B) Relationship between shoot dry weight, UV-B and CO_2 exposures in tomato (*Lycopersicon esculentum* Miller).[24] (C) Relationship between shoot dry weight, UV-B and CO_2 exposures in aster (*Aster tripolium* L.).[24] (D) Relationship between pollen tube length, UV-B and O_3 exposures in petunia (*Petunia hybrida* Vilm.).[21] See color insert for color representation.

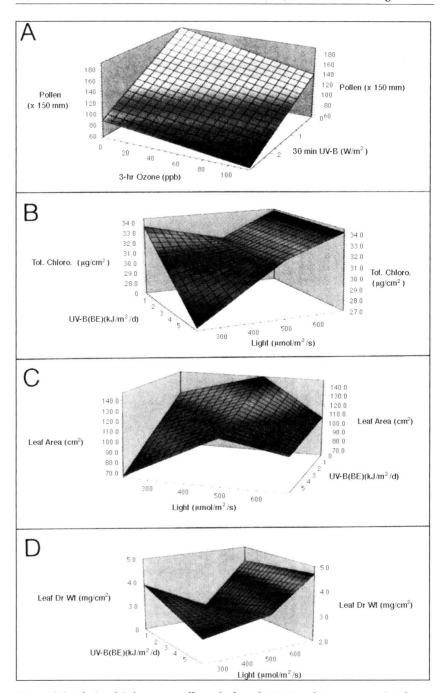

Fig. 7.2. (A) Relationship between pollen tube length, UV-B and O_3 exposures in tobacco (*Nicotiana tabacum* L.).[21] (B) Relationship between leaf total chlorophyll, UV-B and light exposures in bean (*Phaseolus vulgaris* L.).[36] (C) Relationship between leaf area, UV-B and light exposures in bean.[36] (D) Relationship between leaf dry weight, UV-B and light exposures in bean.[36] See color insert for color representation.

there would be evidence of an interaction under the combined exposure. In this context, the search for interactions between simultaneous exposures of two or more stress agents requires the use of modeling.[27]

A good introduction to plant effects modeling for establishing relationships under UV-B exposure was given by Allen.[28] In this context, although intended for climate change effects models for longer time periods (several years), the advice of Hänninen[29] about the need for evaluating model precision and model realism are pertinent for shorter time spans that might be considered in efforts to model UV-B effects on plants. High precision means that the model predictions of the structure and functioning of the system closely agree with observations; high realism means that the model represents causal relationships in the system. Precision can be examined directly; realism cannot and must be inferred not only from high precision in predicting contemporary conditions of climate (e.g., UV-B climate), but also from changed conditions that can only be obtained by rigorous testing of the model under simulated future conditions through experimentation.[29]

The first computer simulation models incorporating ambient UV-B radiation effects on crop growth can be expected to be deterministic models, if for no other reason than due to the scarcity of data to define the definitive quantitative relationships. A statistical model, called the Additive Main Effects and Multiplicative Interaction (AMMI), was designed to enable statistical analysis of unreplicated (deterministic) simulation model output.[30] Part of the procedure is to determine if interactions exist or not so that, if they do, then their output can be eliminated for the purpose of obtaining additive (meta) models that are used for estimating the variance of error in the simulation model. It is then used to compare the simulation output against other sources of data. This could be a useful way for evaluating the existence of interactive effects among multiple environmental factors including UV-B, on crop growth and production responses. The risk is that this approach describes interactions among factors using statistical rather than plant physiological explanations.[31] A useful comparison is to contrast the approach of Willers et al[30] with that of Yan and Wallace.[31]

Deciduous and evergreen dwarf shrubs and mosses were studied under the influence of elevated CO_2 and increased UV-B at Abisko, Sweden.[32] An interaction was observed between CO_2 and UV-B which in combination reduced the growth of one of the deciduous dwarf shrub species. Unfortunately, no detailed statistical analysis was given. Because of the high latitude distribution of this species, it is unknown whether the results have any bearing on deciduous dwarf shrub crops grown elsewhere in temperate latitudes, but it does offer evidence of interactive effects from the two climatic factors.

A recent set of experiments with maize and sunflower (*Helianthus annuus* L.) seedlings exposed to two levels of UV-B, temperature and CO_2 would appear to be aimed at revealing interactions between those environmental factors and various growth parameters. With three factors, plus the response variate, $2^4-1 = 15$, and ignoring examinations of each of the four variates alone (without in combination), 15 - 4 = 11, indicates that eleven treatments should have been used. Unfortunately, only three treatments (plus the control case) were used: increased UV-B; increased temperature and UV-B; and finally, increased UV-B, temperature and CO_2. In such a situation, it is not possible to compare the required treatment effects against each other, because two-thirds (8/11) of the needed treatments were not included. Such studies, while appearing to possibly reveal whether or not there were interactive effects, actually cannot do so, because an insufficient amount of data were collected. In this context, future research should pay attention to the adequacy of the experimental design.

UV-B-O_3 interactions

Pollen from tobacco (*Nicotiana tabacum* L., cv. Bel-W3) and petunia (*Petunia hybrida* Vilm., cv. White Cascade) were exposed for 30 minutes to UV-B at 3 W m^{-2} and/or for 3 hours to ozone at 120 ppb, alone and in sequence (UV-B, and then O_3), followed by an 18 hour period in white light only.[21] In the U.S., the National Ambient Air Quality Standard for O_3 during 1996 was 120 ppb, not to be exceeded by more than one hour annually. Pollen tube length in both species was reduced by UV-B alone (more for petunia, Fig. 7.1D, than for tobacco), and by O_3 alone (more for tobacco, Fig. 7.2A, than for petunia). But, the reduction of pollen tube growth was greatest for both plant species when they were exposed to the sequence of UV-B and then O_3. Feder and Shrier[21] concluded that the joint effects appeared to be additive. In a mathematical sense, this does not appear to be quite true. For tobacco, the joint effect was negative (adverse) and less-than-additive, but with a larger growth reduction in the sequential exposure than under either stress alone. For petunia, the joint effect was also negative and almost additive for the growth reduction, if the response value was combined from either stress alone. These findings might be relevant to the floricultural economics of the European countries, specifically The Netherlands. This is also true for developing African countries, where tobacco cultivation is important.

Specific leaf weight (thickness) of 21-day-old seedlings of snap bean (*Phaseolus vulgaris* L., cv. Bush Blue Lake 290) grown in a growth chamber, showed a statistically significant synergistic increase (+34.9%) when grown under 11.7 kJ m^{-2} day^{-1} plant effective UV-B radiation (simulating a 20% stratospheric ozone depletion), followed by a 3-hour exposure to O_3 at 0.25 ppm.[33] There was no statistically significant change if exposed to O_3 only. But, there was a statistically significant increase of +21.5% if the plants were grown with the elevated UV-B only and no O_3. By contrast, the change in percent dry weight of leaves was statistically significant (-11.3%), but a less-than-additive effect compared to the interaction of a 21-day elevated UV-B, followed by the O_3 exposure. The O_3 exposure alone was not significant. The statistically significant response to only increased UV-B irradiance was considerably higher (-19.5%) than the ameliorative response to both stresses in the exposure sequence used. Statistically significant effects were also found for leaf area, dry and fresh weights, but only as a response to UV-B alone. Photosynthetic rates and stomatal conductance were found to decrease in a statistically significant manner only from the 3-hour O_3 exposure and no interaction was observed with UV-B. Two problems preclude these results from being used for planning in agricultural production: (1) the photosynthetic light level used was too low to represent most field conditions, and (2) the study was limited to only the first 21 days of plant growth.

In one field study initially designed to examine possible interactive effects on growth and production of three soybean cultivars to ozone and UV-B, results showed *no effects* from UV-B under levels simulating a 35% column ozone reduction.[34] No interactions could be found in plant responses to ozone and UV-B exposure. But, the surprisingly contradictory observations of the authors, relative to much of the other similar literature, produced interesting conclusions: many of the studies produced faulty results of a negative effect from: (1) not monitoring ground-level UV-B radiation during the experiments, and therefore (2) relying on a popular mathematical model to estimate ground-level UV-B which inherently overpredicts UV-B levels, and (3) failing to account for changes in UV-B levels as a result of seasonal and climatic changes. It should be concluded that if the inputs to an observed system are not adequately and accurately measured, then the search for interactions can become an exercise in futility.

UV-B-visible light interactions

Regarding the interaction between UV-B radiation and visible light, it is known that plants grown in high visible light have a tendency to show growth responses that are less sensitive to increases in UV-B.[35] Bean (*Phaseolus vulgaris* L., cv. Stella) seedlings were grown in a growth chamber for three weeks under three visible light (700, 500 and 300 μmol m^{-2} s^{-1}), and two plant effective UV-B levels (0.0, and 6.17 kJ m^{-2} day^{-1}).[36] A generalized plant action spectrum was used to compute exposure dose and response. At Lund, Sweden, the higher UV-B value was considered to be at the level that would result on a cloudless day with no measurable aerosols (filter for radiation) and from a 5% reduction in stratospheric ozone during midJuly. The authors appeared to have used the low light and no UV-B situation as a control or reference. It might have been better to use the higher (700 mol μm^{-2} s^{-1}) light level and no UV-B as the control against which resulting changes could be compared and the nature of the joint effects identified. The following analyses are based on that assumption. The 700 μmol m^{-2} s^{-1} light level is not as high as full sunlight at midday at Lund, but it is closer than 300 μmol m^{-2} s^{-1}. However, no test results for determining the statistical significance of apparent differences were given. Among the several bean plant response variables examined, six vegetative growth parameters responded to differences in visible light levels and UV-B (Figs. 7.2B-D and 7.3A-C). Four of these six effects appeared to be negatively synergistic: average total chlorophyll, average leaf area, average leaf dry weight and average leaf fresh weight. When related to the high visible light and no UV-B, all of those growth responses declined more under the combination of decreased light (to 300 μmol m^{-2} s^{-1}) and increased UV-B, compared to either stress alone. The decrease in average leaf thickness, from high light and no UV-B to low light and elevated UV-B, was negative and was nearly additive. Apparently there was little interaction between light and UV-B with regard to this plant

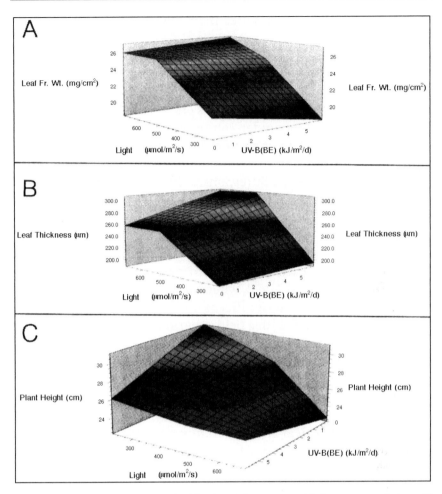

Fig. 7.3. (A) Relationship between leaf fresh weight, UV-B and light exposures in bean.[36] (B) Relationship between leaf thickness, UV-B and light exposures in bean.[36] (C) Relationship between plant height, UV-B and light exposures in bean.[36] See color insert for color representation.

response. Increased UV-B dominated the decrease in leaf thickness and the average plant height change was positive, but was much less than an additive effect.

The bean plants were able to grow taller when limited only by a lack of adequate visible light, than when also exposed to increased UV-B. It is unfortunate that the control or reference value used for UV-B was 0.0, instead of a more realistic value such as 5.6 kJ m^{-2} day^{-1} which represents the ambient UV-B level. Therefore, the results are artificial and this precludes their consideration for the field conditions.

UV-B-moisture interactions

In a growth chamber experiment, cow pea (*Vigna unguiculata* L.) seedlings showed significantly less plant height retardation, fresh weight and dry weight, under conditions of *both* drought and increased UV-B exposure, than to either stress alone that by themselves led to greater retardations.[37] Here, it would appear that each stress factor triggered plant responses that prevented the effects of the other stress from becoming more severe.

In terms of plant responses to increased UV-B under drought stress, either water or nutrient stress has been observed to make many crops less sensitive to UV-B.[38] In one study, field grown soybeans showed reduced leaf area, total plant dry weight and net photosynthesis under increased UV-B, when water was not limiting.[39] But, under drought stress, no significant UV-B effects were found. During May through October in a field experiment, soybean plants were grown under either adequate moisture, or an experimentally induced drought stress, combined with an ambient plant effective UV-B (8.5 kJ m^{-2}) or an elevated UV-B level (13.6 kJ m^{-2}); the latter considered to approximate a 25% stratospheric ozone reduction at Beltsville, MD, U.S.[40] Although there were statistically significant differences in plant dry weight, leaf area and number of pods for the treatment with elevated plant effective UV-B and adequate moisture, under the combination of elevated UV-B and water stress, only the drought effect was statistically significant. There was no interactive effect between decreased moisture and increased UV-B.

So far only a limited amount of evidence indicates a possible interaction between drought stress and effects from experimentally increased UV-B. With respect to interactions between drier climate and potential future increases in UV-B, despite possible claims to the contrary, water stress appeared to lessen the effect of increased UV-B on stomatal resistance during the midday period in cucumber, while increased UV-B stimulated stomatal resistance in radish (*Raphanus sativus* L.).[35,41] In another study, under recovery from water stress and an increase in DNA-effective UV-B to simulate a 23% reduction in stratospheric O_3, 30-day-old soybean seedlings showed an additive, negative effect on net photosynthesis, while 60-day-old plants showed a less-than-additive negative effect to both stress factors.[35,42] Total dry weight response of the 30-day-old plants appeared to represent an additive negative effect from both stress parameters. For the 60-day-old plants, total dry weight response was only due to the water stress and not from the elevated UV-B level. Soybean is a crop of importance in North America in terms of its worldwide production.

UV-B-nutrient interactions

With regard to plant nutrient stress (P deficiency) and UV-B, net photosynthesis and growth of greenhouse-grown soybean plants were unaffected by effective daily exposure of 11.5 kJ m^{-2} UV-B.[43] The interaction of increased UV-B with phosphorus stress resulted in an increase in the UV-B absorbing compounds in the leaf tissue. In lettuce (*Lactuca sativa* L.), reduction in net photosynthesis increased, but percent dry weight reduction decreased, as nutrient availability was increased under elevated UV-B radiation.[35] Another study suggests that an increase in UV-B could lead to stress from iron deficiency and leaf chlorosis in cotton (*Gossypium hirsutum* L.).[44]

It has been stated that future research should consider analyzing the effects of combinations of environmental parameters such as CO_2 enrichment, rise in air temperature, increased UV-B, ozone, SO_2, and other contaminants on long-term plant response under field conditions.[45] Fragmented and patchy information that is available suggests the possibility of multiple stress interactive effects on pea, bean, lettuce, rice, tomato, tobacco and cotton. However, interactive effects of altered levels of UV-B, CO_2, O_3, temperature, water and nutrients on growth and production of subtropical crops such as olives (*Olea europaea* L.), citrus and figs (*Ficus carica* L.) have not been examined.

Many crop plants are annual dicots. Potted plants of a wild dicot of the Asteraceae, *Dimorphotheca pluvialis* Moench, were exposed to various levels of UV-B in a greenhouse experiment under high and low nutrient conditions.[46] With elevated UV-B, some plant properties increased with high nutrient availability, while others increased under low nutrient conditions. On the whole, the authors concluded that success of regeneration and survival of seedlings would be reduced under increased UV-B and low soil fertility.

A growth chamber experiment was conducted with cow pea seedlings (*Vigna unguiculata* L.), with two different levels of available potassium, under two different UV-B levels.[47] Negative interactive effects were observed through decreases in shoot height, fresh weight and dry weight, with leaf area being much greater under both stress treatments compared to those under either the reduced potassium availability or the increased UV-B.

The effects of nitrogen deficiency and increased UV-B radiation on the growth of rye (*Secale cereale* L.) were studied in growth chambers.[48] Increased UV-B and decreased nitrogen availability decreased most production characteristics in an additive manner, without an apparent interaction between UV-B and nitrogen. The exception was that the root/shoot weight ratio was decreased by an interaction between increased UV-B and decreased availability of nitrogen.

CO_2, Temperature, Moisture and Nutrient Availability

While this chapter is focused primarily on the interactions of UV-B with other stress agents and the resulting crop responses, methodologically, it is important to look at how researchers have investigated interactions, aside from increased UV-B. One general assessment states that potential primary production will increase under increased CO_2 and future global warming, because elevated CO_2 levels will increase light, water and nitrogen use efficiency.[49] Elevated levels of atmospheric CO_2 can decrease plant sensitivity to stress such as drought, extreme temperatures and photoinhibition.[50] Elevated CO_2 could also affect plant-herbivore interactions, nutrient cycling and decomposition of organic matter, although studies on these topics are rare, as they relate to actual field conditions. Some agricultural crops are perennials while most others are annuals. Some annuals are determinate—once they switch from vegetative to reproductive growth they cannot switch back, but can only complete their seed production. Indeterminate annuals are more flexible, they continue their vegetative growth along with their reproductive phase and can switch back and forth during the growing season. Also, these species continue seed production until they are killed, for example by an early frost. Many annual crops, although determinate, are harvested while continuing their vegetative growth. However, in response to environmental stress determinate annuals appear to be the most vulnerable, especially during their stages of seed production and seed germination. Next in order in their vulnerability might be the more flexible indeterminate annuals, and lastly, the perennials.[51] However, knowledge of the response of a plant species or cultivar to two or more stress inducing factors singly, does not allow the prediction of the overall integrated response of that plant, when exposures to these factors are in a simultaneous or sequential pattern. Furthermore, the development of methods to study the impacts of multiple stress factors on plants to determine additive, synergistic, or antagonistic effects, is a daunting challenge.[51]

CO_2-light-temperature interactions

The interactive effects of CO_2, light and temperature on the photosynthetic process have been among the most studied. The gross photosynthesis for a whole soybean leaf at CO_2 saturation was modeled as a synergistic interaction between incident photosynthetic photon flux density and leaf temperature.[52] The nature of the interaction with increasing CO_2 would appear to be relative to a given frame of reference of a particular CO_2 concentration, since there exists an intercellular CO_2 saturation level. The increase of CO_2 to 2.3 x ambient (800 ppm) prior to flower expansion in orange (*Citrus sinensis* L.) at suboptimal temperatures was shown to increase subsequent fruit set and thereby, decrease susceptibility to cold stress.[53] This could have positive implications for Latin American and Caribbean countries, the world's leading producers of orange.

An ameliorative interaction between increased CO_2 and low temperature has been observed with okra *(Abelmoschus esculentus* L.).[54] While okra produces fruit under the current ambient levels of CO_2, (~350-360 ppm), and 32°C/26°C daytime/nighttime temperatures, it will perish without fruiting, if grown at lower temperatures of 20°C /14°C. But, the CO_2 level raised to 450 ppm will counteract the chilling stress, and the plant will produce some fruit, although not as many as at the higher temperatures. At the low temperature regime, but with the CO_2 level raised to 1000 ppm, okra will outproduce current warm weather fruit yield. Thus, it is possible that increases in ambient CO_2 might allow the crop to be grown in colder geographic areas in the future.[50]

In the context of the interactions between elevated CO_2 and increased air temperature, in growth chamber experiments with two annual weeds, *Abutilon theophrasti* Medik. (C3) and *Amaranthus retroflexus* L. (C4), *Abutilon* showed no interactive response in biomass accumulation over an 80-day growth period, while *Amaranthus* did.[55] Carbon dioxide levels were raised from 400 to 700 ppm, and the photoperiodic air temperature was increased from 28°C to 38°C.

Although independent increases in CO_2 or air temperature resulted in biomass increases in *Amaranthus* during the 80-day period, when the plants were grown under simultaneous increases in both variables, the data showed an antagonistic, negative interaction and a depression in biomass accumulation. It is unknown whether a more realistic increase in air temperature of, for example, +5°C would show the same increase or whether this interaction would be seen in a field environment and without the sudden, large increase in the CO_2 level.

CO_2-O_3 interactions

A cotton growth and yield model was applied to five cotton growing areas in the U.S. over two decades, from the early 1960s through the mid-1980s.[56] The original model included plant growth as affected by weather variables, but not CO_2 or ground-level ozone concentrations. Revised versions of the model incorporated at first, the two decadal annual CO_2 concentrations, and subsequently, the model included the annual CO_2, plus the average summertime O_3 concentrations and their effects on the growth functions. The most realistic results were obtained when the model was applied to data relevant for Fresno, California, U.S. The inclusion of the 23-year increase in ambient CO_2 levels showed an insignificantly small positive increase in the 23-year mean cotton (lint) yield, and this was explained to be a result of several years of inadequate application of soil nitrogen. The study concluded that with adequate nitrogen, elevated CO_2 probably would have increased lint yields by 10%. The inclusion of 23 years of summertime surface mean O_3 concentrations, along with the increased CO_2, showed a 17% decrease in the 23-year mean yield of lint. While these results for long-term yield of lint indicate a 10% increase from increasing CO_2, and a 17% decrease from simultaneous increases in both CO_2 and O_3, it is not possible to determine the nature of the interaction between CO_2 and O_3, because no simulation was performed using only weather variables and elevated O_3. For example, if O_3 alone would have decreased mean yields by 27%, then the interaction with +10% yield from CO_2 would have been in the opposite direction, resulting in a net decrease of 17%. However, if O_3 alone would have decreased mean yields by considerably less than 27%, e.g., only 10%, then the interaction would have been antagonistic, because the combined effect of -17%, is much greater than the +10% from CO_2 and the -10% from O_3.

Balls et al[57,58] conducted a chamber study to evaluate the interactions between ozone exposure, and microclimatic variables such as relative humidity, light (PAR), vapor pressure deficit (VPD) and air temperature, leading to visible foliar injury on clover (*Trifolium subterraneum* L.). The authors used artificial neural networks to examine the relative importance of input parameters. Accumulated O_3 exposure over a threshold of 40 ppb (AOT40) on a daily basis was found to be most important with regard to visible foliar injury levels, followed in order by relative humidity, leaf age, PAR, VPD and temperature. By itself, this method does not appear to be very satisfactory, because no multivariate mathematical equations or response surface plots originated from this approach. As a preliminary first step in an interaction analysis, it might be useful, however, in sorting out priorities for other subsequent multivariate statistical analyses.

CO_2-nutrient interactions

On a theoretical basis, two interactions have been proposed between nitrogen and CO_2 requirements for plant growth between species, as a means to identify which plants are excluded or favored by the environment, if those growth resources change in the future.[59] According to this concept, each plant species or cultivar has a minimum requirement of available nitrogen to survive. This requirement will be reduced with increased ambient CO_2. In the first theory, if ambient CO_2 is not regarded as a depletable resource, but as simply another environmental factor, then the rate of decline of minimum N requirement as a function of CO_2 for each plant species can be compared. The plant species with the lowest rate of decline in minimum N requirement to change in CO_2 levels will compete best, and presumably be most productive at the present levels of ambient CO_2. As CO_2 levels increase in the future, the plant species that will become more favored could be those with the highest rates of decline in minimum N requirement.[59] In the second theory, CO_2 is a depletable, but necessary plant nutrient. With increasing ambient CO_2 levels, plant species or cultivars that are better CO_2, but poorer N competitors are replaced by other plant species or cultivars that are poorer CO_2, but better N competitors. So far, there is no empirical or experimental evidence to evaluate the validity of either views.

Under apparent ambient (~330 mbar or ppm) and elevated CO_2 (~640 mbar or ppm) concentrations in glasshouses, with full sunlight and optimal water availability, cotton (*Gossypium hirsutum* L., a C3 plant) and corn (*Zea mays* L., a C4 plant) seedlings were grown with nitrogen levels varying from 0.6 mM to 24 mM NO_3^-.[60] The control or reference case can be considered to be the lower nitrogen level and ambient CO_2, because growers often apply nitrogen fertilizer, and the CO_2 levels are expected to continue to increase in the future. After 40 days of growth, changes in cotton dry weight and leaf area per plant were found to increase as a synergistic interaction between increased levels of nitrogen and CO_2. Total N, as a percent of dry weight, showed a positive effect, but it was less than additive for the combination of increased nitrogen and CO_2. After a 30-day growth, change of dry weight in corn was positive, but was only additive. Chlorophyll content, however, showed a positive synergistic interaction between the two growth resources. Although stimulated by increases in total N and leaf area, percent dry weight in corn did not show any interactive effect between N availability and increased CO_2. These results might be relevant to North America, the leading producer of corn in the world, and to cotton production in the developing countries of Asia.

Other nutrient interactions

In the North American western shortgrass prairie, a combined increase in the availability of water and nitrogen during the last two years of a 5-year field experiment, produced synergistic increases in the above-ground biomass for warm season grasses and forbs, compared to experiments with increased water or nitrogen availability individually.[61] Cool season grasses and forbs did not show such an effect.

Glasshouse experiments have been performed to determine the nature of the interactions between manure or compost (nutrient availability), non-nutrient soil structural amendments (moisture availability) and the vegetative yields of tomato and wheat seedlings.[62] Actual values for moisture availability were not given directly, but could be implied by combinations of amendments that changed the soil structure and thereby the availability of moisture and aeration of plant roots. One of the amendments was polyacrylamide, a long-chain organic polymer that binds clay particles together similar to natural organic matter. Vegetative growth of wheat was found to respond positively in some cases, but as less-than-additive effects between supplements of poultry manure/compost (the nutrient) and additions of both lignite and dry polyacrylamide (soil structural change). One combination of these additions displayed an additive joint effect on wheat yield.[62]

Tomato seedling growth responded only additively from an interaction between supplemented poultry manure/compost and polyacrylamide, applied as dry granules to the soil.[62] In another experiment, the interaction between polyacrylamide applied as a solution and dairy manure, a negative interaction reduced the dry weight of tomato seedlings, far below the individual effects of the two factors. A possible explanation is that the increased soil aeration from structural improvements led to an increased decomposition rate of the dairy manure high in nitrogen and phosphorus content, and that could have resulted in plant toxicity. Combinations of two and three different soil amendments without nutrient additions from manure showed a reverse, but positive synergistic interaction on tomato seedling growth. The experimental alleviations that were applied did not identify interactions between soil moisture uptake rates and additions of specific nutrient elements for the two crops. However, the study did show how various effects, such as positive-additive, negative-additive, positive-synergistic, and negative-synergistic interactions between multiple stress factors can be evaluated mathematically. All too often such explicit methods are not used in the evaluations of climate change effects.

Moisture-temperature interactions

In the future if the dry Mediterranean climates are altered during the summers by increasing moisture levels in the autumn and early winter, such a change could inhibit the onset of cold hardiness in some olive cultivars of southern Europe. The same moisture-temperature interaction might be found for fruit and nut trees of other European countries. The management practice of irrigating olive orchards resulted in this type of interactive effect, with an exceptionally cold winter in 1990, in the olive producing areas of California, U.S.[63] Production impacts depend upon whether the inhibition of cold hardiness coincides with a subsequent high production year in those cold-sensitive cultivars. Other added costs, such as the pruning of dead branches, or whole tree replacements can be incurred regardless of the production cycle.

Plant Disease and Multiple Stress Factors

Plants that are under temperature and/or drought stress can become more susceptible to disease,[2] since plant resources needed for damage repair are reduced under such conditions. A positive feedback interaction is established and the diseased plant is less able to tolerate other sources of added stress, such as increases in temperature, moisture deficit, elevated UV-B radiation or air pollutants. Changes in the productivity of agriculture are not necessarily expected to be continuous and linear at a level that is easily monitored. Synergistic ecological changes can be latent for a period and then, can suddenly become obvious in a nonlinear manner. An analogy might be made to a person descending a convex mountainside. At first, upon leaving the ridge line, the descent can be almost imperceptible, but after a time the descent can become nearly vertical if one does not change course. A likely result of any UV-B increase, if global warming also occurs, the two factors combined may amplify the adverse effects of each.[2]

In considering interactions, Manning and von Tiedemann[64] identified nine mechanisms by which changes in plant-pathogen interactions can occur: (a) induction of foliar necrotic lesions; (b) change in leaf cuticle integrity; (c) enhancement of plant tissue membrane permeability and leakage of electrolytes; (d) production of antibiotics; (e) changes in leaf surface microflora; (f) change in leaf stomatal function; (g) change in foliage canopy structure; (h) changes in carbon partitioning among plant organs; and (i) earlier plant mortality (senescence). Chapter 5 discusses how changes in CO_2, O_3, and UV-B can dynamically function, resulting in some of these changes in plant-pathogen interactions.

Paul et al[65] in their review, consider the possible effects of increased UV-B on interactions between host plants and various consumer organisms, such as insect herbivores, phytopathogens and decomposers across trophic levels in the ecosystem. Elevated UV-B can affect the host prior to or after the exposure to the consumer, to directly influence either or both components. Because the diversity of hosts and consumers is so broad, the scant available information is generally not quantitative, but qualitative in its nature, (i.e., an interaction has, or has not been found between UV-B, the host and the consumer). This is a different from the concept of interaction than we have been using in a quantitative sense to indicate change resulting from two or more causes that is greater or lesser the additive change from the individual factors.

Increased plant vulnerability to pathogens and insect pests and thus, increased occurrences of disease and pest epidemics could occur especially in regions where winters become milder.[66] Forty-day-old, chamber grown sugar beet (*Beta vulgaris* L.) seedlings, showed no statistically significant change in storage root biomass when grown without increased UV-B, but with a fungal leaf spot disease (*Cercospora beticola* Sacc.). This was also the case when plants were grown without the fungal disease, but under 6.91 kJ m^{-2} day^{-1} plant-effective UV-B radiation.[67] This UV-B level simulated a 9% reduction in stratospheric ozone on a day with clear sky (June 15), and presumably with no measurable atmospheric aerosol content in Lund, Sweden. However, when the plants were grown with the leaf spot disease and increased UV-B, a negative synergistic change in storage root biomass of -23% was observed. If this interaction could be verified in ambient field experiments over a full growth season, it could have potential importance for future agriculture in Germany, and the European countries (Belgium, Germany, Ireland, Netherlands and United Kingdom) where sugar beet is an important crop.

Concerning the interactions between disease and water availability to plants, there is evidence that for the fast spreading annual weed, the groundsel (*Senecio vulgaris* L.), infection with the rust (*Puccinia lagenophorae*) increased the diversity of the plant size, lowered the plant water status and made them susceptible to other stresses such as drought.[68] The smaller the size of the plant, the greater the combined effect. It has been suggested that pathogens and other stress factors generally exert additive, negative effects, wherein each effect decreases the plasticity of the plant or variability of the response. Fifteen days after inoculation, the joint effects of rust infection and simultaneous increase in water stress on groundsel could cause a *negatively synergistic* response in net photosynthesis on a per plant basis and on the leaf area. By contrast, seven weeks after inoculation, the joint effects of rust infection, with a simultaneous decrease in nutrient availability and increase in nutrient stress, showed a compensatory additive, interactive effect on the leaf area ratio and the root-shoot ratio.[68] The disease alone increased the leaf area ratio and decreased the root-shoot ratio. By comparison, a decrease in nutrient availability alone decreased the leaf area ratio and increased the root-shoot ratio. However, both stress factors combined, produced a plant response that was not very different from the control (no rust and adequate nutrients). These results showed that seven weeks after inoculation, the increase in groundsel specific root length due to the rust disease, along with depletion of nutrients, was far less-than-additive, but still greater than the effect from either variable alone.

Insect Herbivory and Multiple Stress Factors

Some plant species show decreases in leaf nitrogen content in response to increased CO_2. For those plants, insect herbivory might eventually decline because of the decrease of forage quality.[59] However, management practices of nitrogen fertilizer application might augment any declines in herbivory. Increased CO_2, decreased soil nitrogen and increased air temperature will all decrease plant tissue nitrogen, making it less palatable or desirable to insect herbivores.[69] Decreased soil nitrogen, reduction in cloudiness (increased periods of clear skies), decreases in moisture availability and increases in temperatures tend to increase the level of plant secondary metabolites. This also makes the plant tissue less palatable or desirable to insect herbivores.[69] On the opposite side, increased soil nitrogen, decreased temperature, increased cloud cover and increased moisture would increase plant nitrogen content, decrease levels of plant secondary metabolites and make the plant more palatable or desirable for insect herbivores. Changes in temperature and moisture might have the largest effect, but the quantitative nature of the simultaneous interaction of all these factors (i.e., how they balance each other for a given insect pest on a given crop) is unknown at present.

There are laboratory studies that suggest that if cloudiness were to increase, leading to lower light levels in environments rich in nitrogen, then beetles might become greater pests and consume five times more willow (*Salix dasyclados* L.) foliage compared to the conditions of high light levels, regardless of the foliar nitrogen content.[70] In contrast, levels of natural insect repellents and toxins (furanocoumarins) were found to increase by nearly 50% in field grown wild parsnip (*Pastinaca sativa* L.) with ambient UV, in comparison to plants grown in sunlight, but without UV.[71] Inadequate characterization of the spectral UV irradiance precludes the interpretation of what this might mean for crops, if UV-B levels increase in the future.

Crop Growth Prediction Models

Many published computer simulation models exist for the important crops across the world, especially for those in the temperate latitude. A few models have been published for crop competition with weeds. Almost all of these are designed to simulate crop growth and production in response to solar radiation, temperature, moisture and in some cases, nutrient levels, with only a few such models incorporating a direct CO_2 response. We know of no crop growth and production simulation models that are currently designed to address changes in UV-B, even though one plant response model was designed to study the effects of increased UV-B and light on photosynthesis in wheat and wild oat grass (*Avena fatua* L.).[72] Modifications of the current crop simulation models to respond to changes in CO_2 and UV-B should be undertaken to explore the level of uncertainty in the assessments of the effects of UV-B and other climatic changes on crops.

A conceptual dynamic, compartmental, plant carbon-growth model has been described that could integrate the combined interactions of day-to-day changes in photosynthetically active solar radiation, ambient CO_2, moisture and nutrient availability, and stress effects from air pollutants on the mass balance of carbohydrate production and consumption by plant organs (roots, stems, leaves and fruits).[73] Known and hypothetical interactions of UV-B with these factors, and the consequential joint effects of such interactions, would also have to be incorporated into the structure of such a model.

References

1. Forseth IN. Plant response to multiple environmental stresses: Implications for climatic change and biodiversity. In: Reaka-Kudla ML, Wilson DE, Wilson EO, eds. Biodiversity II—Understanding and Protecting Our Biological Resources. Washington, DC: Joseph Henry Press, 1997:187-196.
2. Myers N. Synergisms: Joint effects of climate change and other forms of habitat destruction. In: Peters RL, Lovejoy TE, eds. Global Warming and Biological Diversity. New Haven, CT: Yale University Press, 1992:344-354.
3. German Bundestag. Ozone depletion, changes in UV-B radiation and their effects. In: Protecting the Earth—A Status Report with Recommendations for a New Energy Policy. Bonn, Germany: German Bundestag, 1992:571-616.

4. Scientific Committee on Problems of the Environment (SCOPE). Effects of Increased Ultraviolet Radiation on Biological Systems. Paris, France: SCOPE, 1992.
5. van de Staaij JWH, Lenssen GM, Stroetenga M et al. The combined effect of elevated CO_2 and UV-B radiation on growth characteristics of Elymus athericus. Vegetatio 1993; 104/105:433-439.
6. Billick I, Case TJ. Higher order interactions in ecological communities: What are they and how can they be detected? Ecology 1994; 75(6):1529-1543.
7. Bornman JF, Teramura AH. 1993. Effects of ultraviolet-B radiation on terrestrial plants. In: Young AR, Björn LO, Moan J et al, eds. Environmental UV Photobiology. New York: Plenum Press, 1993:427-471.
8. Caldwell MM, Teramura AH, Tevini M et al. Effects of increased solar ultraviolet radiation on terrestrial plants. Ambio 1995; 24(3):166-173.
9. Kreeb KH, Chen T. Combination effects of water and salt stress on growth, hydration and pigment composition in wheat (Triticum aestivum L.): A mathematical modeling approach. In: Esser G, Overdieck D, eds. Modern Ecology—Basic and Applied Aspects. New York: Elsevier, 1991:215-231.
10. Kimball BA. Carbon Dioxide and Agricultural Yield: An Assemblage and Analysis of 770 Prior Observations. Report 14. Phoenix, AZ: USDA, U.S. Water Conservation Laboratory, 1983.
11. Kimball BA. Carbon dioxide and agricultural yield: An assemblage and analysis of 430 prior observations. Agron J 1983; 75:779-788.
12. Kimball BA. Influence of elevated CO_2 on crop yield. In: Enoch HZ, Kimball BA, eds. Carbon Dioxide Enrichment of Greenhouse Crops, Vol. II—Physiology, Yield, and Economics. Boca Raton, FL: CRC Press, 1986:105-115.
13. Cure JD. Carbon dioxide doubling response: A crop survey. In: Strain BR, Cure JD, eds. Direct Effects of Increasing Carbon Dioxide on Vegetation. DOE/ER-0238. Washington, DC: U.S. Dept. of Energy, 1985:99-116.
14. Cure JD, Acock B. Crop responses to carbon dioxide doubling: A literature survey. Agr Forest Meteorol 1986; 38:127-145.
15. Rogers HH, Bingham GE, Cure JD et al. Responses of selected plant species to elevated carbon dioxide in the field. J Environ Qual 1983; 12(4):569-574.
16. Rogers HH, Thomas JF, Bingham GE. Response of agronomic and forest species to elevated atmospheric carbon dioxide. Science 1983; 220:428-429.
17. Krupa SV, Kickert RN. The Greenhouse Effect: Impacts of ultraviolet-B (UV-B) radiation, carbon dioxide (CO_2), and ozone (O_3) on vegetation. Environ Pollut 1989; 61(4):263-393.

18. Krupa SV, Kickert RN. The Greenhouse Effect: The impacts of carbon dioxide (CO_2), ultraviolet-B (UV-B) radiation and ozone (O_3) on vegetation. Vegetatio 1993; 104/105:223-238.
19. Runeckles VC, Krupa SV. The impact of UV-B and ozone on terrestrial vegetation. Environ Pollut 1994; 83:191-213.
20. Miller JE, Pursley WA. Effects of ozone and UV-B radiation on growth, UV-B absorbing pigments and antioxidants in soybean. Plant Physiol 1990; 93(S):101.
21. Feder WA, Shrier R. Combination of UV-B and ozone reduces pollen tube growth more than either stress alone. Environ Exper Bot 1990; 30(4):451-454.
22. Mulchi CL, Slaughter L, Saleem M et al. Growth and physiological characteristics of soybean in open-top chambers in response to ozone and increased atmospheric CO_2. Agr Ecosyst Environ 1992; 38:107-118.
23. Rozema J, van de Staay J, Costa V et al. A comparison of the growth, photosynthesis and transpiration of wheat and maize in response to enhanced ultraviolet-B radiation. In: Abrol YP, Govindjee, Wattal PN et al, eds. Impact of Global Climatic Changes on Photosynthesis and Plant Productivity. New Delhi, India: Oxford & Ibh Publ. Co. PVT. Ltd., 1991:163-174.
24. Rozema J, Lenssen GM, van de Staaij JWM. The combined effect of increased atmospheric CO_2 and UV-B radiation on some agricultural and salt marsh species. In: Goudriaan J, van Keulen H, van Laar HH, eds. The Greenhouse Effect and Primary Productivity in European Agro-Ecosystems. Wageningen, The Netherlands: Pudoc, 1990:68-71.
25. Rozema J, Lenssen GM, van de Staaij JWM et al. Effects of UV-B radiation on terrestrial plants and ecosystems: Interaction with CO_2 enrichment. Plant Ecol 1997; 128(1-2):182-191.
26. Teramura AH, Sullivan JH, Ziska LH. Interaction of elevated ultraviolet-B radiation and CO_2 on productivity and photosynthesis characteristics in wheat, rice and soybean. Plant Physiol 1990; 94:470-475.
27. Rozema J. Plant responses to atmospheric carbon dioxide enrichment: Interactions with some soil and atmospheric conditions. Vegetatio 1993; 104/105:173-190.
28. Allen Jr LH. Plant effects and modeling responses to UV-B radiation. In: Biggs RH, Joyner MEB, eds. Stratospheric Ozone Depletion/UV-B Radiation in the Biosphere. Berlin: Springer-Verlag, 1994:303-310.
29. Hänninen H. Assessing ecological implications of climatic change: Can we rely on our simulation models? Climatic Change 1995; 31:1-4.
30. Willers JL, Wagner TL, Sequeira RA et al. Analysis of deterministic simulation models using methods applicable to two-way treatment structures without replication. Agron J 1995; 87(3):478-492.

31. Yan W, Wallace DH. A model of photoperiod X temperature interaction effects on plant development. Critical Rev Plant Sci 1996; 15(1):63-96.
32. Gehrke C, Johanson U, Gwynn-Jones D et al. Effects of enhanced ultraviolet-B radiation on terrestrial subarctic ecosystems and implications for interactions with increased atmospheric CO_2. Ecol Bull 1996; 45:192-203.
33. Agrawal M, Agrawal SB, Krizek DT et al. Physiological and morphological responses of snapbean plants to ozone stress as influenced by pretreatment with UV-B radiation. In: Abrol YP, Govindjee, Wattal PN et al, eds. Impact of Global Climatic Changes on Photosynthesis and Plant Productivity. New Delhi, India: Oxford & Ibh Publ. Co. PVT. Ltd., 1991:133-146.
34. Fiscus EL, Miller JE, Booker FL. Is UV-B a hazard to soybean photosynthesis and yield? Results of an ozone—UV-B interaction study and model predictions. In: Biggs RH, Joyner MEB, eds. Stratospheric Ozone Depletion/UV-B Radiation in the Biosphere. Berlin: Springer-Verlag, 1994:135-147.
35. Teramura AH. Interaction between UV-B radiation and other stresses in plants. In: Worrest RC, Caldwell MM, eds. Stratospheric Ozone Reduction, Solar Ultraviolet Radiation and Plant Life; Workshop on the Impact of Solar Ultraviolet Radiation upon Terrestrial Ecosystems: 1. Agricultural Crops, Bad Windsheim, 27-30 September 1983, West Germany. New York: Springer-Verlag, 1986:327-343.
36. Cen Y-P, Bornman JF. The response of bean plants to UV-B radiation under different irradiances of background visible light. J Exp Bot 1990; 41(232):1489-1495.
37. Balakumar T, Hani Babu Vincent V, Paliwal K. On the interaction of UV-B radiation (280-315 nm) with water stress in crop plants. Physiol Plant 1993; 87:217-222.
38. Tevini M, Teramura AH. UV-B effects on terrestrial plants. Photochem Photobiol 1989; 50(4):479-487.
39. Murali NS, Teramura AH. Effectiveness of UV-B radiation on the growth and physiology of field-grown soybean modified by water stress. Photochem Photobiol 1986; 44(2):215-220.
40. Sullivan JH, Teramura AH. Field study of the interaction between solar ultraviolet-B radiation and drought on photosynthesis and growth in soybean. Plant Physiol 1990; 92:141-146.
41. Teramura AH, Tevini M, Iwanzik W. Effects of ultraviolet-B irradiation on plants during mild water stress. I. Effects on diurnal stomatal resistance. Physiol Plant 1983; 57(2):175-180.
42. Teramura AH, Perry MC, Lydon J et al. Effects of ultraviolet-B radiation on plants during mild water stress. III. Effects on photosynthetic recovery and growth in soybean. Physiol Plant 1984; 60(4):484-492.

43. Murali NS, Teramura AH. Insensitivity of soybean photosynthesis to ultraviolet-B radiation under phosphorus deficiency. J Plant Nutr 1987; 10(5):501-516.
44. Jolley VD, Brown JC, Pushnik JC et al. Influences of ultraviolet (UV)-blue light radiation on the growth of cotton. I. Effect on iron nutrition and iron stress response. J Plant Nutr 1987; 10(3):333-352.
45. Rozema J, Lenssen GM, Arp WJ. Global change, the impact of the Greenhouse Effect (atmospheric CO_2 enrichment) and the increased UV-B radiation on terrestrial plants. In: Rozema J, Verkleij JAC, eds. Ecological Responses to Environmental Stresses. Dordrecht, The Netherlands: Kluwer Academic Publ., 1991:220-231.
46. Musil CF, Wand SJE. Differential stimulation of an arid-environment winter ephemeral *Dimorphotheca pluvialis* (L.) Moench by ultraviolet-B radiation under nutrient limitation. Plant, Cell Environ 1994; 17:245-255.
47. Premkumar A, Kulandaivelu G. Influence of ultraviolet-B enhanced solar radiation on growth and photosynthesis of potassium deficient cowpea seedlings. Photosynthetica 1996; 32(4):521-528.
48. Deckmyn G, Impens I. Combined effects of enhanced UV-B radiation and nitrogen deficiency on the growth, composition and photosynthesis of rye (*Secale cereale*). Plant Ecol 1997; 128(1-2):235-240.
49. van Keulen H. The impact of the Greenhouse Effect on factors limiting primary production. In: Goudriaan J, van Keulen H, van Laar HH, eds. The Greenhouse Effect and Primary Productivity in European Agro-Ecosystems. Wageningen, The Netherlands: Pudoc, 1990:62-63.
50. Hogan KP, Smith AP, Ziska LH. Potential effects of elevated CO_2 and changes in temperature on tropical plants. Plant Cell Environ 1991; 14(8):763-778.
51. Bazzaz FA, Morse SR. Annual plants: Potential responses to multiple stresses. In: Mooney HA, Winner WE, Pell EJ, eds. Response of Plants to Multiple Stresses. New York: Academic Press, 1991:283-305.
52. Harley PC, Weber JA, Gates DM. Interactive effects of light, leaf temperature, CO_2, and O_2 on photosynthesis in soybean. Planta 1985; 165:249-263.
53. Downton WS, Grant WJR, Loveys BR. Carbon dioxide enrichment increases yield of Valencia orange. Aust J Plant Physiol 1987; 14:493-501.
54. Sionit N, Strain BR, Beckman HA. Environmental controls on the growth and yield of okra. I. Effects of temperature and of CO_2 enrichment at cool temperature. Crop Sci 1981; 21:885-888.
55. Coleman JS, Bazzaz FA. Effects of CO_2 and temperature on growth and resource use of co-occurring C3 and C4 annuals. Ecology 1992; 73(4):1244-1259.

56. Reddy VR, Baker DN, McKinion JM. Analysis of effects of atmospheric carbon dioxide and ozone on cotton yield trends. J Environ Qual 1989; 18:427-432.
57. Balls GR, Palmerbrown D, Cobb AH et al. Towards unravelling the complex interactions between microclimate, ozone dose and ozone injury in clover. Water, Air, Soil Pollut 1995; 85(3):1467-1472.
58. Balls GR, Sanders GE, Palmerbrown D. The use of artificial neural networks to model plant environment interactions. Pesticide Sci 1996; 46(3):280-282.
59. Tilman D. Carbon dioxide limitation and potential direct effects of its accumulation on plant communities. In: Kareiva PM, Kingsolver JG, Huey RB, eds. Biotic Interactions and Global Change. Sunderland, MA: Sinauer Associates Inc., 1993:333-346.
60. Wong SC. Elevated atmospheric partial pressure of CO_2 and plant growth. I. Interactions of nitrogen nutrition and photosynthetic capacity in C3 and C4 plants. Oecologia 1979; 44:68-74.
61. Lauenroth WK, Dodd JL. The effects of water- and nitrogen-induced stresses on plant community structure in a semiarid grassland. Ecology 1978; 36:211-222.
62. Wallace A, Wallace GA. Additive and synergistic effects on plant growth from polymers and organic matter applied to soil simultaneously. Soil Sci 1986; 141:334-342.
63. Denney JO, Martin GC, Kammereck R et al. Lessons from a recordbreaking freeze... some olives show damage; many, coldhardiness. Calif Agric 1993; 47(1):1-12.
64. Manning WJ, von Tiedemann A. Climate change: Potential effects of increased atmospheric carbon dioxide (CO_2), ozone (O_3) and ultraviolet-B (UV-B) radiation on plant diseases. Environ Pollut 1995; 88:219-245.
65. Paul ND, Rasanayagam S, Moody SA et al. The role of interactions between trophic levels in determining the effects of UV-B on terrestrial ecosystems. Plant Ecol 1997; 128(1-2):296-308.
66. Harte J, Torn M, Jensen D. The nature and consequences of indirect linkages between climate change and biological diversity. In: Peters RL, Lovejoy TE, eds. Global Warming and Biological Diversity. New Haven, CT: Yale University Press, 1992:325-343.
67. Panagopoulos I, Bornman JF, Björn LO. Response of sugar beet plants to ultraviolet-B (280-320 nm) radiation and Cercospora leaf spot disease. Physiol Plant 1992; 84:140-145.
68. Ayres PG. Growth responses induced by pathogens and other stresses. In: Mooney HA, Winner WE, Pell EJ, eds. Response of Plants to Multiple Stresses. New York: Academic Press, 1991:227-248.
69. Ayres MP. Plant defense, herbivory and climatic change. In: Kareiva PM, Kingsolver JG, Huey RB, eds. Biotic Interactions and Global Change. Sunderland, MA: Sinauer Associates Inc., 1993:75-94.

70. Larsson S, Wiren A, Lundgren L et al. Effects of light and nutrient stress on leaf phenolic chemistry in *Salix dasyclados* and susceptibility to *Galerucella lineola* (Coleoptera). Oikos 1986; 47:205-210.
71. Zangerl AR, Berenbaum, MR. Furanocoumarins in wild parsnip: Effect of photosynthetically active radiations, ultraviolet light and nutrients. Ecology 1987; 68:516-520.
72. Ryel RJ, Barnes PW, Beyschlag W et al. Plant competition for light analyzed with a multispecies canopy model. I. Model development and influence of enhanced UV-B conditions on photosynthesis in mixed wheat and wild oat canopies. Oecologia 1990; 82:304-310.
73. Jäger H-J, Grünhage L, Dämmgen U et al. Future research directions and data requirements for developing ambient ozone guidelines or standards for agroecosystems. Environ Pollut 1991; 70:131-141.

CHAPTER 8

World Agricultural Production

Introduction

There are several levels of evidence which might be considered as most to least reliable in the philosophy of science applied to solving environmental problems: (a) long-term field monitoring data; (b) model responses with minimal field data; (c) model implications for expected or possible field responses, but with no actual data on model responses; and (d) inferences based on analogous reasoning or theory-based hypotheses with little existing evidence to evaluate.

In the case of best possible knowledge, the analysis and evaluation would be based extensively on field monitoring data for all stress and response variables. However, this level is seldom achieved; rather, it is a matter of how far short of this ideal situation, we view an environmental problem. In relatively neglected areas of science, the ideal level of evidence is often not possible to achieve on an extensive basis, but only as a few scattered and individual case studies. In strictly physical systems, such as the global spatial variation of UV-B radiation, a database will eventually be established. When the scope is enlarged to examine the effects of UV-B on global agricultural production, then political and socio-economic dimensions are brought into consideration. It then becomes even more difficult to foresee obtaining ideal knowledge via long-term field monitoring as the basis for impact assessment.

Failing the availability of the highest order of evidence, scientific evaluations are then forced to the next, lower level of evidence that is based on the use of quantitative, often computer simulation-based, models of the system(s) under consideration. Some modelers frequently contend that a model is even better than any single set of field monitoring data, because the model, if sufficiently and successfully field-tested, will contain the behavior of many different response

Elevated Ultraviolet (UV)-B Radiation and Agriculture, by Sagar V. Krupa, Ronald N. Kickert and Hans-Jürgen Jäger.
© 1998 Springer-Verlag and Landes Bioscience.

systems, whereas any set of field data are only representations of one particular manifestation. However, when the state of the science is at a level where we are forced to perform almost all of our reasoning on the basis of using mathematical simulation models, then it is advisable that we rely on model responses from field tests. Unfortunately, in many cases, responses are published for models that had not been field tested.

In cases where model responses have been described in the literature in only a minimal way, when compared to the objectives of application, although the structure of the model might be well documented, one is forced to infer model implications of what the responses might be. This is quite a primitive level of evidence and reasoning, based on much subjectivity.

Lastly, in areas of science where we have no field monitoring data, and no models in any formal sense other than as vague concepts, we are forced to rely on deduced inferences or on analogous reasoning. This might also be considered as theory-based hypothesis development, with little if any existing evidence to evaluate the adequacy of the hypothesis. It is satisfactory for planning a research program, but has the highest level of uncertainty, and is therefore not ideal for any type of policy analysis and decision making. Unfortunately, this appears to be the level of evidence on which the following analysis must be based.

There have been a number of published evaluations of the effects of possible future climate change on agriculture,[1-7] but these evaluations are rather simple and only consider the effects of spatial changes in air temperature and precipitation on crop production. Such simplified analyses can lead to questionable conclusions such as "...climate change will have only mild impacts on average global agricultural output and may even improve temperate agricultural production."[7] These types of efforts have been criticized as depending heavily on the major (temperate) cereals which are not especially the dominant crops of developing countries. "Further investigations to include crops such as bean (*Phaseolus vulgaris* L.), root crops, sugarcane (*Saccharum officinarum* L.) and various fruit crops would give a more comprehensive picture of the likely situation in tropical regions... ."[4]

In the early 1990s, a 3-year study was performed to project the effects of climate change on world food production.[8] Climate change was defined as only changes in ambient CO_2, air temperature, precipitation and soil moisture. The assessment time period was from the year 1990 to 2060. At the global level, the study used climate change scenarios resulting from three well known General Circulation Models (GCM), GISS (Goddard Institute of Space Studies, U.S.), GFDL (Geophysical Fluid Dynamics Laboratory, U.S.) and UKMO (United Kingdom Meteorological Office). For crop production at the global level, this study used the Basic Linked System of National Agricultural Models. Components included in this analysis were agricultural production, population, economics and political policy. Results varied not only geographically for a given GCM, but also varied depending upon which GCM was used as the basis for climate change input to the crop production assessment. Depending upon the GCM, world crop production changed between approximately +1 to -10%.

A recent review of prospects for change and threats to global agriculture under possible future climate change, examined on a continental basis, the change during the last three decades of the human population, the production of cereals, tubers, and pulses, arable land, and during the last decade, the amount of fertilizer used and land irrigated.[9] Notable observations by continent include: (a) in Africa, cereal production declined in comparison to the rate of change in population, and tuber crop production was relatively high; (b) in Asia, the rate of increasing cereal production has been the highest, but pulses as a source of protein increased only slightly, while this continent has the highest percentage of irrigated land; (c) in South America, the increase in arable land, i.e., land use converted to agriculture, is highest; (d) Europe, North America and Central America had the lowest rate of increase in population, while their rate of crop production increased, making them highly favorable as sources for export, with the exception of the decline in tuber production in Europe during the last three decades; and (e) in Oceania, there has been a dramatic increase in the growth rate of pulses. After considering present and possible future changes in atmospheric CO_2, UV-B, tropospheric ozone, air temperature, evapotranspiration, humidity, rainfall, cloudiness, radiation, and soil properties, the authors concluded that, overall "...the relationship between climate change and agriculture is still very much a matter of conjecture, with many uncertainties..."[9]

At a recent International Conference on Ozone Depletion and Ultraviolet Radiation: Preparing for the Impacts, "Protecting the Food

Supply Working Group" barely mentioned the need to investigate the impacts of changing UV-B on Third World food supplies that have suffered in the past from inadequate crop management practices for several decades.[10]

It has been suggested that scientists in global ecotoxicology should work toward the goal of making a first estimate of the potential significance of contaminants (and UV-B) as agents of global change.[11] The author identified the occurrence of increasing UV-B as a unique aspect of global change, one that has continental to global spatial distribution, with subtle effects, a relatively constant frequency of exposure, and is likely to last for many decades to greater than a century. We do not fully concur with this statement. To assess the possible impact of increased, geographically patchy ground-level UV-B radiation on world agriculture, it is necessary to compare crops with some cultivars for which sensitivity to UV-B is known (chapter 3), against the production levels of various crops across the world. Such an assessment is hampered by the inadequacy of the data on ground-level UV-B and its effects on many crops, especially in the tropics. Furthermore, there can be extreme intra-species (cultivar, varietal) variability in the sensitivity of a given crop species to UV-B.

An ideal assessment would compare the weeds of the world (chapter 5) and their competitive abilities within the various crops, in terms of changes in future crop productivity under increased UV-B. Depending upon the nature of the crop, and the sociological culture of the end consumers, marketable productivity can have both quantitative and qualitative attributes. For example, in the U.S., consumers do not purchase cosmetically blemished apples (*Malus pumila* Miller), even if the size and the weight of the apples are not affected. Other countries, with other cultural values, do not particularly respond in the same manner. In addition, an ideal assessment would also account for any changes in crop production (loss) through indirect effects of UV-B on diseases and insect pests (chapter 6). We acknowledge that there are also crop losses due to vertebrate pests. Furthermore, the ideal assessment would account for the interactions between altered climate and its direct effects on crops and indirect effects from weed competition, diseases and insects (chapter 7). While we can recognize the kinds of information that are needed, the present state of knowledge does not enable such an ideally integrated, comprehensive assessment of the likely future impacts of changes in UV-B radiation on crops. Here, our attention is directed at both intensive, as well as extensive, agricultural regions, the latter being defined by Walker[3] as regions with little or no cultivation of crops for commercial purposes (subsistence level cropping), in contrast to intensive management for high productivity.

According to Hale,[12] at the present time there are no data on UV-B effects that address crop production and changes over time, at the farm level, prior to food processing. While we agree with this observation, scientists should make the initial, best approximation, given the limited knowledge available, even if it proves to be inaccurate in the future. Planners at the national level are sometimes adept at deducing how much it will cost a country for example, to phase out the production of CFCs (chlorofluorocarbons), but are completely remiss at not accounting for the amount of economic savings that will result elsewhere in the society by doing so (a balanced cost-benefit analysis) or not including an economic analysis of the costs to society (crop losses) by not changing existing policies and industrial production practices.[13]

During 1990, in the U.S. crop income totaled $80.4 billion[14] and we can compare this to the loss figures given by Hale.[12] The latter does not give a year of reference, but the magnitudes should be roughly comparable. Hale reports the following crop losses in the U.S. due to: (a) diseases ~$9.1 billion, (b) insect pests ~$7.7 billion, (c) weeds ~$6.2 billion, and (d) losses to tropospheric ozone (not resulting from decreases in stratospheric ozone and increases in ground-level UV-B) ~$3.1-$3.2 billion. These losses total ~$26.1 billion from the potential crop income, or approximately a 21.6% reduction. Estimates provided in Chatterjee[13] for the quantification of effects on the U.S. economy until year 2075 from global implementation of the Montreal Protocol, show the benefits in billions of dollars, as of 1985. This is for UV-B and for additional tropospheric ozone-induced crop losses and can be deduced as $2.19 billion and $1.22 billion, respectively (total $3.41 billion) for the year 1990, assuming a constant annual inflation rate of 3% over the 90-year period (1985- 2075). This is 4.2% of the 1990 total crop income in the U.S. ($80.4 billion). To the extent that the crop income in the U.S. during the 1990s has not already experienced a net reduction, any future additional net losses from increased UV-B could boost this total percentage of crop loss to 21.6% for diseases and insect pests, and to 25-30% for weeds. One investigator has suggested that any economic impact from crop production losses as a result of future increases in UV-B might be greatly diminished, because the changes would occur slowly and growers would make changes in their crop production practices to mitigate against such losses.[15] This view might represent a cultural bias that has greater possibility in developed countries practicing commercial agriculture, than in developing countries where subsistence farming does not provide a large choice of alternatives. Worldwide, there are geographic differences in states of economic development and in cultural values. These aspects must be considered when attempting to assess future prospects on a global scale.

Nonetheless, societies and their political representatives will not be able to, or will want to wait until all crops, weeds, diseases and insect pests have been experimentally screened for their sensitivity to UV-B. Those crops that are known to be growth and production sensitive to increased UV-B radiation can be viewed across the entire complex of crops for which production data are available at the global scale. In this manner, at least a first-order estimation can be made as to which regions of the world might be particularly vulnerable to increased UV-B radiation. The questions that need to be answered are: (1) Which crops could be impacted? (2) Which world regions dominate the production of those crops? (3) What could be the amount of production change for those crops in those regions? and (4) What would be the cross-societal impacts of such changes in world crop production? We present a discussion of the first two questions by continent or region, based on 1996 crop production data from the United Nations' Food and Agriculture Organisation (U.N. FAO.) To address the third question requires more research, and in spite of that, we discuss some ways in which the fourth question may be addressed.

Current Agricultural Production

By their very nature, grain and spice crops have an advantage over root, tuber, vegetable and fruit crops, since they have a relatively low moisture content and can be stored for longer periods, with small risk of perishing. For millennia, these features have allowed them to represent the internationally traded crop commodities. This feature has also allowed careful accounting of production statistics for many decades for grains as opposed to other crops. The importance of world agricultural production by various crops can be classified as: (1) international trade where export is a source of national income and import is a national cost (wheat, *Triticum aestivum* L., rice, *Oryza sativa* L. and corn, *Zea mays* L.); (2) raw material resource for manufacturing value added products (sugar beet, *Beta vulgaris* L., rapeseed, *Brassica napus* L., olive, *Olea europaea* L., tobacco, *Nicotiana tabacum* L., grapes, *Vitis vinifera* L., cotton, *Gossypium hirsutum* L. and forage for livestock); and (3) directly consumable local and national food resources (potato, *Solanum tuberosum* L., fruits and vegetables). The lack of availability of crop production data often follows this same order—world crop production data are mostly available for internationally traded cereal grains and are relatively scarce for specific fruit and vegetable crops.

World crop production statistics for 1996, from the FAO[16] database, are shown by major regions in Table 8.1, and the scientific names of individual crops are presented in Appendix 1. These major regions are: North America (the U.S. and Canada), Latin America and the Caribbean, developing countries of Africa, Western Europe transition markets (including Eastern European and former Soviet block countries), developing countries of Asia, and Oceania, respectively, and include the nations shown in Appendix 2. The developed countries of Asia consist of Israel and Japan. For comparative purposes, a column of total world production by crop category is shown in Table 8.1. However, this amount does not necessarily match the sums across regions for a given crop, due to variability in definitions of the regions, and does not include every country in the world. In some cases, records do not pertain to a specific crop type and they are listed in a general category with the suffix "nes." With the exception of those crops that are only grown in one region of the world, Table 8.1 shows the production figures and the region that led in maximum production of a given crop in 1996.

Sugarcane is the leading crop with highest production figures and Latin America and the Caribbean are the leading regions for that crop. On a worldwide basis, for the major crops, recent production figures show wheat, maize and rice are the highest (Table 8.1). World yields for potato, sugar beet, and grasses for forage and silage follow in that order.

Possible Future Changes in Crop Production

In a recent review including implications for agriculture, Caldwell et al[17] state that, due to the relatively few field studies that have been conducted, a quantitative prediction of the potential

Table 8.1. 1996 World crop production by continent or sub-continent or region (metric tonnes) from the FAO database[a]

Crop	North America	Latin America & Caribbean	Developing Africa	Western Europe	Transition Markets (incl. E. Europe)	Developing Asia	Developed Asia	Oceania	World
Wheat	92,594,000	24,233,570	19,897,740	99,437,210	90,159,180	231,100,400	700,000	21,167,850	582,042,900
Rice	7,771,000	19,934,210	15,461,930	2,696,600	1,390,893	506,086,000	13,000,000	968,240	567,311,900
Barley	24,552,000	1,880,602	8,306,982	54,916,470	39,272,080	20,045,810	227,000	6,417,025	155,794,000
Maize	243,364,000	68,785,810	36,259,350	34,141,380	29,895,230	154,442,600	2,900	481,753	577,724,000
Rye	551,020	51,076	25,400	6,376,367	16,090,360	940,030	—	21,000	24,058,250
Oats	6,627,000	737,930	229,400	7,305,293	13,155,760	961,464	2,700	1,613,790	30,669,340
Millet	180,000	44,150	12,343,070	500	735,490	13,941,210	700	27,000	27,285,120
Sorghum	20,396,500	8,840,607	20,571,350	572,400	104,230	16,152,580	300	1,555,960	68,673,920
Buckwheat	55,000	53,000	—	23,000	1,057,409	1,012,000	18,000	—	2,218,709
Triticale	—	—	—	3,029,815	2,216,760	1,100,000	—	537,000	6,883,575
Quinoa	—	37,181	—	—	—	—	—	—	37,181
Fonio	—	—	191,000	—	—	—	—	—	191,000
Canary seed	165,000	21,500	3,500	300	—	2,550	—	3,000	195,850
Mixed grain	—	—	560,000	1,057,850	—	20,500	—	—	5,264,130
Cereals nes[b]	—	10,740	2,092,500	11,461	3,625,780	46,750	—	6,500	2,242,551
Potatoes	26,323,340	13,410,780	6,803,107	50,680,660	74,600	81,387,940	3,645,000	1,403,905	289,069,100
Sweet potatoes	617,110	1,819,119	7,389,011	57,173	103,945,700	122,161,000	1,187,900	560,337	133,847,800
Cassava	—	31,418,680	83,225,380	—	—	46,263,600	—	188,956	161,096,600
Yautia (coco yam)	—	249,451	—	—	—	—	—	—	249,451
Taro (coco yam)	2,680	30,217	3,578,645	—	—	1,529,393	260,000	331,200	5,732,135
Yams	—	847,058	31,683,170	1,100	—	32,000	210,000	288,420	33,061,750

World Agricultural Production

Crop									
Roots & tubers nes[b]	—	735,768	2,200,230	6,500	—	877,551	55,000	331,520	4,206,569
Sugarcane	26,511,600	529,223,500	58,568,670	169,000	—	503,895,200	1,610,000	44,951,840	1,187,442,000
Sugar beets	25,130,900	3,360,273	3,891,537	111,959,600	74,913,560	30,610,520	3,800,000	—	253,666,400
Sugar crops nes[b]						510,000			510,000
Beans, dry	1,431,700	5,357,509	2,028,514	144,496	330,146	8,667,095	130,000	34,000	18,183,460
Broad beans, dry	18,500	150,062	910,000	290,170	22,063	1,996,200	250	120,000	3,507,245
Peas, dry	1,587,740	104,013	360,120	3,415,846	3,405,312	1,848,909	1,400	424,490	11,150,750
Chickpeas	—	173,448	268,950	118,014	1,780	8,035,263	5,500	304,000	8,906,955
Cow peas, dry	1,920	30,860	2,298,600	—	32,200	78,670	1,000	3,000	2,452,050
Pigeon peas	—	46,100	199,300	—	—	3,172,179	—	—	3,417,579
Lentils	481,080	34,352	53,000	38,953	31,212	2,125,212	50	51,000	2,814,859
Bambara beans	—	—	51,000	—	—	—	—	—	51,000
Vetches	—	—	89,730	134,133	465,490	228,100	950	41,000	959,403
Lupins	—	20,805	9,000	40,920	76,200	1,420	—	1,230,000	1,384,345
Pulses nes[b]	—	29,445	1,007,875	567,400	189,275	1,939,937	780	5,127	3,739,839
Brazil nuts	—	43,400	5,200	—	—	—	—	—	48,600
Cashew nuts	—	191,505	196,400	—	—	334,050	—	—	721,955
Chestnuts	—	—	270	138,187	15,210	292,000	35,000	—	480,667
Almonds	401,500	3,310	135,150	404,053	22,771	253,785	3,100	8,500	1,232,169
Walnuts	186,000	41,900	3,000	90,242	214,387	498,095	—	70	1,033,694
Pistachios	60,000	—	1,325	6,000	—	374,855	—	—	442,180
Kolanuts	—	—	308,500	—	—	—	—	—	308,500
Hazelnuts	17,240	—	140	122,004	8,018	464,500	—	—	611,902
Areca nuts (betel)	—	—	100	—	—	513,916	—	—	514,016
Nuts nes[b]	140,000	71,109	90,680	9,827	12,200	94,420	1,500	27,114	451,350
Soybeans	67,090,000	39,278,320	565,652	1,047,400	790,434	20,910,520	120,000	73,000	129,959,300
Groundnuts in shell	1,653,200	885,988	6,103,476	5,445	11,850	20,340,220	49,300	48,930	29,228,400
Coconuts	—	—	1,753,071	—	—	37,831,260	—	1,941,000	44,673,170
Oil palm fruit	—	4,932,000	14,384,000	—	—	69,981,000	—	1,220,000	90,517,000

Table 8.1. continued

Crop	North America	Latin America & Caribbean	Developing Africa	Western Europe	Transition Markets (incl. E. Europe)	Developing Asia	Developed Asia	Oceania	World
Palm kernels	—	391,825	922,560	—	—	3,745,000	—	70,000	5,129,385
Palm oil	—	964,815	1,827,181	—	—	14,007,780	—	276,000	17,075,780
Olives	150,600	155,439	1,967,000	7,700,310	63,972	2,204,460	36,500	1,000	12,279,280
Karite nuts (sheanuts)	—	—	662,500	—	—	—	—	—	662,500
Castor beans	—	67,236	36,430	—	1,710	1,301,114	—	—	1,411,490
Sunflower seeds	1,686,900	5,800,593	420,868	3,975,800	8,676,434	4,093,289	22,400	87,000	25,483,280
Rapeseed	5,255,690	93,466	183,000	7,047,922	1,593,607	15,490,390	1,000	588,800	30,253,870
Tung nuts	—	243,000	4,200	—	—	404,929	—	—	652,129
Safflower seed	190,810	109,000	35,000	350	6,000	470,890	70	27,000	839,120
Sesame seed	—	147,162	556,995	250	110	1,864,355	350	—	2,569,222
Melonseed	—	—	589,100	2,500	—	45,700	—	—	637,300
Mustard seed	382,220	—	—	8,557	79,880	130,600	—	200	601,457
Poppy seed	—	—	—	12,357	25,900	7,300	—	—	45,557
Tallowtree seeds	—	—	—	—	—	795,000	—	—	795,000
Vegetable tallow	—	—	—	—	—	119,250	—	—	119,250
Stillingia oil	—	—	—	—	—	119,250	—	—	119,250
Kapok fruit	—	—	—	—	—	447,300	—	—	447,300
Kapokseed in shell	—	—	—	—	—	335,219	—	—	335,219
Seed cotton	10,722,000	4,519,855	4,462,896	1,397,000	6,220,477	26,881,590	113,168	959,000	55,400,290

World Agricultural Production

Cottonseed	6,596,000	2,568,292	2,669,398	755,000	3,883,210	17,726,390	70,418	595,000	34,939,500
Linseed	883,690	102,247	56,600	208,167	125,410	932,345	—	15,150	2,323,609
Hempseed	—	1,100	—	3,847	1,963	28,700	—	—	35,610
Oil seeds nes[b]	—	10,100	200,837	235,900	28,839	1,041,688	20,000	100	1,548,464
Cabbages	2,033,000	853,297	701,426	3,647,726	12,987,450	24,718,240	2,770,631	119,370	48,061,160
Artichokes	49,000	95,290	91,100	908,212	—	19,610	4,300	—	1,167,512
Asparagus	95,559	162,418	36,453	205,787	3,500	2,510,312	10,160	14,800	3,041,989
Lettuce	3,643,000	337,364	153,260	3,193,283	39,300	5,373,804	571,400	142,580	13,483,990
Spinach	226,520	23,850	72,514	506,743	17,300	5,168,328	362,000	6,300	6,383,555
Tomatoes	12,305,000	8,370,051	9,130,279	12,440,410	8,080,136	32,128,600	1,256,551	515,220	84,702,640
Cauliflower	342,000	156,386	150,400	2,170,072	347,286	9,226,075	175,000	110,000	12,707,220
Pumpkins; squash; gourds	38,000	1,272,312	889,300	793,700	1,310,000	4,923,330	276,000	171,210	9,953,852
Cucumbers & gherkins	1,086,800	458,660	376,300	1,645,492	2,439,369	16,247,600	911,400	22,360	23,202,980
Eggplants	28,710	59,580	548,655	488,500	45,000	10,241,670	513,900	2,220	11,928,230
Chillies+peppers; green	626,300	1,238,661	1,807,910	1,494,792	1,052,300	7,668,918	224,200	25,261	14,138,540
Onions+shallots; green	—	740,699	452,910	181,791	16,000	1,171,337	540,000	170,020	3,272,757
Onions, dry	2,934,000	2,496,112	2,023,695	3,037,047	3,793,780	20,278,390	1,377,864	246,032	36,495,420
Garlic	232,000	277,733	196,330	325,555	274,593	9,056,538	8,500	1,400	10,372,650
Leeks & other alliac. veg.	—	4,600	660	845,301	1,000	689,372	—	3,200	1,544,133
Beans, green	187,800	153,922	163,200	842,455	121,448	1,984,163	83,200	50,590	3,616,798
Peas, green	1,156,200	192,708	241,240	1,800,004	394,201	1,207,389	55,700	140,000	5,203,442
Broad beans, green	—	190,345	207,100	234,727	1,500	334,830	5,290	3,900	977,692
String beans	880,000	54,818	4,000	310,000	—	156,013	—	244	1,405,075

Table 8.1. continued

Crop	North America	Latin America & Caribbean	Developing Africa	Western Europe	Transition Markets (incl. E. Europe)	Developing Asia	Developed Asia	Oceania	World
Carrots	2,015,000	916,880	836,925	3,488,887	3,417,621	4,914,795	794,845	245,500	16,675,860
Okra	1,350	39,595	945,000	—	—	363,810	—	1,200	1,350,955
Green corn	4,380,000	481,950	1,117,500	—	120,000	425,820	435,000	299,085	7,529,355
Mushrooms	415,500	—	2,740	807,856	130,751	595,267	75,360	53,754	2,083,228
Chicory roots	—	—	100	317,100	27,700	—	—	—	362,900
Carobs	—	120	26,000	187,194	—	18,000	200	—	231,514
Vegetables fresh nes[b]	1,133,000	4,089,552	10,437,000	9,127,700	6,709,100	156,336,400	3,344,000	429,476	191,868,300
Vegetables canned nes[b]	—	—	—	—	46,150	733,500	—	32,800	812,450
Bananas	6,000	23,531,250	6,631,850	422,798	—	23,759,260	105,700	918,421	55,513,090
Plantains	—	6,343,516	21,717,600	—	—	845,000	—	4,800	28,910,910
Oranges	10,635,000	28,863,300	3,653,204	5,376,984	281,442	9,867,802	516,900	453,446	60,379,220
Tang., mand., clement., satsma	499,860	1,961,945	1,039,850	2,080,219	600	9,055,057	1,329,143	85,259	16,051,930
Lemons & limes	948,000	2,961,053	514,420	1,315,873	—	3,599,309	26,810	37,640	9,463,456
Grapefruit & pomelo	2,465,700	1,050,336	240,490	51,133	—	620,418	404,390	28,970	5,021,075
Citrus fruit nes[b]	—	175,778	2,661,820	19,834	—	931,152	322,500	12,155	4,123,239
Apples	5,217,627	3,444,509	895,350	9,520,366	7,903,847	23,658,330	1,097,300	859,000	53,092,260
Pears	726,500	779,969	241,200	2,952,219	824,688	6,736,919	456,000	195,000	13,122,500
Quinces	—	54,720	32,528	23,863	51,077	163,980	1,500	1,070	329,338

World Agricultural Production

Apricots	73,719	50,130	221,540	544,774	426,772	1,010,922	11,800	42,000	2,421,657
Sour cherries	133,598	—	—	—	481,557	543,191	156,000	—	1,314,346
Cherries	146,500	24,925	5,000	741,252	510,811	378,486	17,300	9,400	1,834,374
Peaches & nectarines	982,805	814,140	214,100	4,138,885	601,971	3,306,877	215,000	95,000	10,503,780
Plums	854,069	264,500	127,530	953,588	2,186,783	2,688,839	149,600	34,000	7,281,909
Stone fruit nes[b]; fresh	—	—	6,350	16,300	13,000	295,400	8,000	1,500	340,550
Strawberries	771,170	138,633	36,600	701,124	440,968	242,746	214,500	13,200	2,562,541
Raspberries	53,400	—	150	64,424	221,374	—	—	1,200	340,548
Gooseberries	—	—	—	92,030	96,313	—	—	30	188,373
Currants	—	—	—	251,303	433,461	—	—	2,520	687,284
Blueberries	137,000	—	50	3,000	16,500	—	—	1,500	158,050
Cranberries	229,470	—	50	—	7,500	—	—	—	237,020
Berries nes[b]	19,300	7,527	2,280	34,195	13,527	236,800	—	4,100	318,229
Grapes	5,083,194	5,806,372	1,320,800	24,718,880	7,304,204	10,457,080	336,065	1,124,500	57,811,100
Watermelons	1,852,000	1,774,712	1,921,950	2,017,346	4,569,069	16,903,490	741,500	80,600	29,910,670
Cantaloupes+other melons	966,680	1,391,784	1,004,570	1,806,700	716,115	9,733,624	490,000	75,475	16,214,950
Figs	38,800	35,850	284,030	207,933	16,886	491,890	550	135	1,077,174
Mangoes	3,630	2,860,192	1,859,350	500	—	14,374,130	15,340	36,812	19,179,140
Avocados	167,800	1,445,445	151,312	53,800	—	124,500	50,000	17,470	2,046,327
Pineapples	315,000	2,893,489	1,791,184	2,000	—	6,693,260	26,000	176,924	12,032,020
Dates	19,140	1,651	1,387,238	8,000	—	2,982,120	11,500	—	4,409,649
Persimmons	—	47,000	—	48,999	—	1,037,440	269,500	2,240	1,405,179
Cashewapple	—	1,250,000	66,000	—	—	—	—	—	1,316,000
Kiwi fruit	28,070	160,000	—	424,542	—	3,405	47,400	204,500	867,917
Papayas	20,410	3,049,423	756,045	—	—	1,868,082	50	16,500	5,729,510
Fruit tropical fresh nes[b]	—	756,227	392,205	32,000	—	6,453,659	200	29,926	7,664,217

Table 8.1. continued

Crop	North America	Latin America & Caribbean	Developing Africa	Western Europe	Transition Markets (incl. E. Europe)	Developing Asia	Developed Asia	Oceania	World
Fruit fresh nes[b]	7,440	782,776	3,883,653	257,800	515,300	20,151,890	14,100	533,195	26,201,160
Straw; husks	—	54,431,000	—	49,839,810	19,887,000	188,200	175,000	200	124,521,200
Maize for forage+silage	80,399,000	20,695,000	—	112,807,300	226,005,100	368,300	5,701,000	315,000	446,290,700
Sorghum for forage+silage	3,952,000	44,189,000	—	314,278	—	584,000	1,844,000	3,700	50,886,980
Ryegrass; forage+silage	—	1,913,000	—	2,144,400	—	—	—	—	4,057,400
Grasses nes[b], forage+silage	90,639,900	7,170,200	—	65,256,200	29,310,340	3,526,000	34,357,000	—	230,259,600
Alfalfa for forage+silage	72,010,000	30,480,830	—	41,952,820	4,930,700	7,840,600	—	1,230,000	158,445,000
Cabbage for fodder	—	—	—	1,390,400	—	—	—	600,000	1,990,400
Pumpkins for fodder	—	—	—	22,000	754,200	—	—	—	776,200
Beets for fodder	—	5,250	—	8,573,360	376,950	100,000	—	2,750	9,058,310
Carrots for fodder	—	—	—	47,300	39,000	—	—	—	86,300
Swedes for fodder	—	—	—	100,000	—	—	—	2,640	102,640
Clover for forage+silage	—	—	46,200,000	18,820,680	681,000	2,137,500	—	—	67,839,180
Leguminous nes[b]; forage+silage	—	—	—	32,642,800	15,506,700	693,000	17,000	—	48,859,500

World Agricultural Production

									Total
Leaves & tops; vines	—	—	—	—	—	19,164,070	—	—	19,164,070
Forage products nes[b]	—	40,000	510,300	16,500,700	—	116,323,500	—	3,222,000	137,796,500
Turnips for fodder	—	—	—	1,614,300	645,200	—	80,000	5,830	2,345,330
Vegetables+roots; fodder	—	—	—	820,500	41,067,300	431,500	5,000	—	42,324,300
Coffee, green	2,360	3,528,668	1,166,828	—	—	1,114,127	—	65,143	5,877,126
Cocoa beans	—	648,771	1,562,901	—	—	452,414	—	35,137	2,699,223
Tea	—	60,594	378,004	40	80,475	2,040,930	90,000	9,000	2,671,018
Mate	—	454,000	200,000	—	—	—	—	—	654,000
Hops	34,500	330	—	42,849	24,862	14,450	1,100	3,300	121,561
Pepper; white/long/black	—	37,756	7,690	—	800	190,705	—	156	237,107
Pimento+allspice	10,000	45,288	391,417	10,500	65,210	1,317,406	360	—	1,851,181
Vanilla	—	378	2,166	—	—	2,445	—	82	5,071
Cinnamon (canella)	—	40	2,380	—	—	63,400	—	—	65,820
Cloves; whole+stems	—	—	23,321	—	—	111,930	—	—	135,251
Nutmeg; mace; cardamons	—	19,041	890	—	—	42,130	—	—	62,061
Anise; badian; fennel	—	3,310	58,360	2,344	8,500	113,605	—	—	186,119
Ginger	4,260	2,425	81,125	—	—	490,318	—	1,995	580,123
Spices nes[b]	—	4,633	26,365	3,000	994	835,157	—	774	870,923
Peppermint	—	8,000	49,000	800	4000	120	1,500	—	63,420
Pyrethrum; dried flowers	—	70	16,700	200	—	—	—	500	17,470
Cotton lint	4,126,000	1,540,485	1,654,354	494,000	1,965,367	9,054,582	42,750	420,000	19,341,040
Flax fibre & tow	—	4,000	12,500	120,140	190,909	251,535	—	—	579,084
Hemp fibre & tow	—	4,000	—	4,442	18,644	76,065	20	—	103,171
Kapok fibre	—	—	—	—	—	110,000	—	—	110,000
Jute	—	3,335	5,100	—	—	2,511,600	—	—	2,520,035

Table 8.1. continued

Crop	North America	Latin America & Caribbean	Developing Africa	Western Europe	Transition Markets (incl. E. Europe)	Developing Asia	Developed Asia	Oceania	World
Jute-like fibres	—	31,185	9,700	—	45,000	449,196	—	—	536,281
Ramie	—	4,969	—	—	—	63,800	—	—	68,769
Sisal	—	192,344	86,030	—	—	16,485	—	—	297,531
Agave fibres nes[b]	—	**62,710**	—	—	—	3,500	—	—	**66,210**
Abaca (manila hemp)	—	19,059	550	—	—	88,600	—	—	108,209
Coir	—	—	—	—	—	169,200	—	2,500	171,700
Fibre crops nes[b]	46,500	80,300	40,140	—	—	**255,400**	—	590	**426,130**
Tobacco leaves	782,070	739,738	436,489	320,618	358,268	3,742,734	69,700	8,026	6,481,537
Natural rubber	—	70,811	268,104	—	—	6,264,266	—	3,700	6,606,881
Natural gums	—	22,850	—	—	—	—	—	—	22,850
Hay non-leguminous	—	—	500,000	—	**88,485,000**	—	—	—	88,985,000
Hay (clover; lucerne; etc.)	—	—	18,300	—	—	1,900,000	—	3,200,000	5,118,300
Hay (unspecified)	—	—	300,000	3,500,000	**58,024,900**	—	210,000	—	62,034,900
Range pasture	—	—	**58,700,000**	1,500,000	—	—	—	—	60,200,000
Improved pasture	—	—	—	6,500,000	—	—	—	—	6,500,000
								TOTAL	7,263,782,082

[a] With the exception of those crops that are only grown in one region of the world, the table shows the production figures and the region that led in maximum production for a given crop in 1996. [b] In some cases, records do not pertain to a specific crop type and therefore, are listed in a general category with the suffix "nes."

consequences for global food production resulting from increased solar UV-B is not possible at present. If the word quantitative is interpreted in the traditional sense, on an interval or ratio scale,[18] then we agree with Caldwell et al.[17] However, at the ordinal scale, based on indications in the literature, a crude first glimpse might be obtained by comparing relative sensitivity rankings.

The regions of maximum production by crop are listed in Table 8.2, together with an indication of the relative sensitivity or tolerance of a crop to changes in UV-B radiation with regard to its biomass accumulation (chapter 3). It should be recognized that some investigators regard much of the previously published results on the response of crop species to UV-B exposure to be flawed, because the artificial exposures were actually much higher than what can reasonably be expected from stratospheric ozone depletion in the future.[19] Although they may be correct, it is not possible at this time to re-evaluate the sensitivities of the various crops mentioned in chapter 3, although they serve as the basis of what is presented in this chapter.

The developing countries in Asia have the greatest diversity of crops and lead the world in the production of those crops. Crop diversity in Asia is followed by Western Europe, then the developing countries of Africa. Oceania, including Australia and New Zealand have the least diversity of crops, although the two countries lead world production of those crops. This variability is significant for international agricultural trade and for the degree of vulnerability of a given region to any future climate changes.

A number of concepts emerge from examination of Tables 8.1 and 8.2. There are a large number of crops for which there is no information on the effects of changes in UV-B on production: (a) sugarcane, the world's leading crop, is sensitive to UV-B and since Latin American and the Caribbean regions with relatively small agricultural diversity are so dependent upon it, those geographic regions could be quite vulnerable to any increases in UV-B; (b) the developing countries of Asia, and the developed countries of Western Europe, lead the world in the production of a number of crops that are sensitive to changes in UV-B; (c) considering the small diversity in major crop production, it appears that the transition market countries, including Eastern Europe, as a region are in an even more vulnerable situation; and (d) North America, with the exception of soybean (*Glycine max* (L.) Merr.) production, does not appear

Table 8.2. World leading crops by region based on FAO data for 1996; Developed Asia is not the top leading producer in any crop and is not listed

UV-B Sensitive	Moderately UV-B Sensitive		UV-B Tolerant			
North America	Latin America & Carribean	Developing Africa	Western Europe	Transition Markets (incl. E. Europe)	Developing Asia	Oceania
Maize	Sugarcane	**Sorghum**	Barley	**Rye**	**Wheat**	Lupins
Canary seed	Plantains	Cassava	Triticale	Oats	Rice	Kiwi fruit
Soybean		Taro (coco yam)	Sugar beets	Mixed grain (cereals nes)[a]	**Millet**	Hay (clover; lucerne; etc)
Mustard seed	Avocados	Yams	Peas, dry	Vetches	Sweet potatoes	—
String beans	Cashewapple	**Roots & tubers nes**[a]	Almonds	Poppy seed	**Beans, dry**	—
Green corn	Papayas	Cow peas, dry	Olives	Buckwheat	Broad beans, dry	—
Grapefruit and pomelo	Straw; husks	Kolanuts	**Artichokes**	**Potatoes**	Chickpeas	—
Blueberries	**Sorghum for forage+silage**	Karite nuts (sheanuts)	Leeks & other alliac. veg.	**Sunflower seeds**	Pigeon peas	—
Cranberries	Coffee, green	Melonseed	Peas, green	Sour cherries	Lentils	—
Grasses nes[a]; forage+silage	Mate	Okra	Mushrooms	Raspberries	Pulses	—
					Carrots	
					Vegetables fresh nes[a]	
					Vegetables canned nes[a]	
					Bananas	
					Tang, mand., clement., satsma	
					Lemons and limes	
					Apples	
					Pears	
					Quinces	
					Apricots	

World Agricultural Production

Alfalfa for forage+silage	Sisal	**Clover for forage+silage**	Chicory roots	Gooseberries	Cashew nuts	Plums
	Agave fibres nes[a]	Cocoa beans	Carobs	Currants	Chestnuts	**Stone fruit nes**[a]**; fresh**
		Peppermint	Cherries	**Maize for forage+silage**	Walnuts	**Berries nes**[a]
		Pyrethrum; dried flowers	Peaches & nectarines	**Pumpkins for fodder**	Pistachios	Grapes
		Range pasture	Strawberries	**Vegetables+ roots; fodder**	**Groundnuts in shell**	**Watermelons**
			Ryegrass; forage+silage	Hay non-leguminous	Coconuts	**Cantaloupes+ other melons**
			Cabbage for fodder	Hay (unspecified)	Oil palm fruit	Figs
			Beets for fodder		Palm kernels	Mangoes
			Carrots for fodder		Palm oil	Pineapples
			Swedes for fodder		Castor beans	Dates
			Leguminous nes[a]**; forage+ silage**		Rapeseed	Persimmons
			Turnips for fodder		Tung nuts	**Fruit tropical fresh nes**[a]
			Hops		Safflower seed	**Fruit fresh nes**[a]
					Sesame seed	**Forage products nes**[a]
					Seed cotton	Tea
					Cottonseed	Pepper; white/long/black

Table 8.2. continued

UV-B Sensitive	Moderately UV-B sensitive		UV-B Tolerant			
North America	Latin America & Carribean	Developing Africa	Western Europe	Transition Markets (incl. E. Europe)	Developing Asia	Oceania
—	—	—	—	—	Linseed	—
—	—	—	—	—	Pimento+ allspice	—
—	—	—	—	—	Vanilla	—
—	—	—	—	—	Hempseed	—
—	—	—	—	—	Oil seeds nes[a]	—
—	—	—	—	—	Cinnamon (canella)	—
—	—	—	—	—	Cloves; whole +stems	—
—	—	—	—	—	Cabbages	—
—	—	—	—	—	Nutmeg; mace; cardamons	—
—	—	—	—	—	Asparagus	—
—	—	—	—	—	Anise; badian; fennel	—
—	—	—	—	—	Lettuce	—
—	—	—	—	—	Spinach	—
—	—	—	—	—	Ginger	—
—	—	—	—	—	Tomatoes	—
—	—	—	—	—	Spices nes[a]	—
—	—	—	—	—	Cauliflower	—
—	—	—	—	—	Cotton lint	—
—	—	—	—	—	Pumpkins; squash; gourds	—
—	—	—	—	—	Flax fibre & tow	—

Cucumbers & gherkins	—	—	—	—	—	—
Eggplants	—	—	—	—	—	—
Chillies+peppers; green	—	—	—	—	—	—
Onions+shallots; green	—	—	—	—	—	—
Onions, dry	—	—	—	—	—	—
Garlic	—	—	—	—	—	—
Beans, green	—	—	—	—	—	—
Broad beans, green	—	—	—	—	—	—

Hemp fibre & tow	—	—	—	—	—	—
Jute	—	—	—	—	—	—
Jute-like fibres	—	—	—	—	—	—
Ramie	—	—	—	—	—	—
Abaca (manila hemp)	—	—	—	—	—	—
Fibre crops nes[a]	—	—	—	—	—	—
Tobacco leaves	—	—	—	—	—	—
Natural rubber	—	—	—	—	—	—

[a] In some cases, records do not pertain to a specific crop type and therefore, are listed in a general category with the suffix "nes."

to be very vulnerable for those crops in which it leads the world production.

It has been claimed that small island countries are especially vulnerable to sea level rise that would result if climate warming occurred, and to increased UV-B from the loss of stratospheric ozone.[13] Such a claim appears to be unsubstantiated for the data for Oceania in Table 8.1, with the exception of sugarcane, barley (*Hordeum vulgare* L.) and leguminous hay production. In comparison, there are simply too few tropical crops for which UV-B sensitivity is known. Any future changes in the production of UV-B-sensitive crops in various regions of the world (Table 8.2), would certainly lead to a realignment of exports and imports of tradable crop commodities, as well as to changes in local nontransportable subsistence farming products. This would mean significant economic and political changes from the present world situation.

Other approaches to evaluating effects on global agriculture are possible. For example, anticipated or hypothesized UV-B induced changes in agricultural productivity can be entered on a national or a regional basis as inputs into a global, multi-sector (environment, natural resources, economics, social, energy, political) computer simulation model and the future consequences throughout the society can be analyzed on a yearly basis. This was done in the context of possible climate (heat and moisture) change effects on agriculture.[20] That study used the global International Futures Simulation (IFS) model.[21] However, additional research is needed in that direction with regard to the possible impacts of increased UV-B.

Comparison of the base case scenario of the IFS (Professional Edition, Version 3.06[21]) from 1992 through 2050 (with no fractional decline in crop yield, Fig. 8.1), with two other simulations during which a gradual decline occurs from 0 in 1992 to -20 and -40% in 2050, shows several responses (Fig. 8.2). Table 1.5 in chapter 1 shows trends in annual UV-B values as "percent per decade." Referring to the changes in plant damage action spectra, if such changes persisted roughly for five decades, one could expect the following changes in UV-B by the year 2050: at 75°N: +54%; at 5°N/S: +9%; and at 75°S: +195%. Using the gradual declines in crop yield of 20 and 40% does not seem unreasonable with the projections, even though we would hope that such a future does not occur. It should be mentioned that this model includes the effects of rising ambient atmospheric carbon dioxide, over the simulated time period.

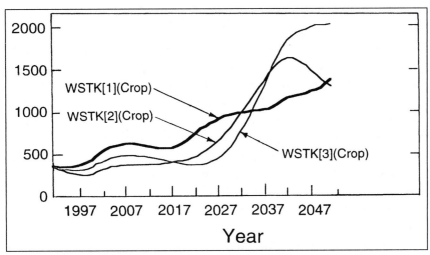

Fig. 8.1. Changes in World Food Stock in the International Futures Simulation[22] (Professional Edition, Version 3.06) from year 1992 through year 2050. WSTK is World Food Stock (MMT) in the Agriculture/Food Sector, [1] the Base case with no imposed change in global crop yield; [2] under a gradual linear 40% decline in global crop yield from 1992 through 2050; [3] under a gradual linear 20% decline in global crop yield from 1992 through 2050.

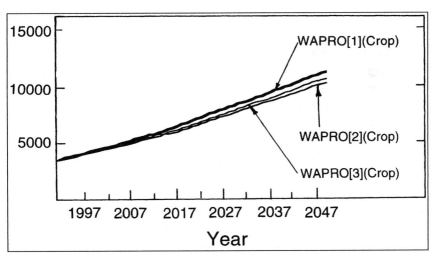

Fig. 8.2. Changes in World Agricultural Crop Production (WAPRO) (MMT) in the International Futures Simulation (Professional Edition, Version 3.06) from year 1992 through year 2050, [1] no change in global crop yield (Base case); [2] gradual decline from 0.0% in 1992 to -40%; [3] and to -20% in agricultural yield at year 2050.

From Figure 8.1, it is evident that the simulated World Food Stock (WSTK), under both a 20 and 40% reduction in yield, leads to a progressively smaller amount each year for the first decade and during part of the next century. In the latter part of the second decade, the simulated World Food Stock falls lower for the 20% decrease in yield, than for the 40% decline. This disparity continues until midway through the second quarter of the next century (circa 2037). At that point, Simulated World Food Stock under a 20% decline in yield rises to the highest level for the first half of the next century. However, the simulated World Food Stock for the 40% decline in yield reaches a peak around year 2040, after which it falls, possibly indicating a threat to the sustainability of the global population.

Under both these gradual declines in crop yield during the next 50 years, annual World Agricultural Crop Production (WAPRO) in the International Futures Simulation is less than the Base case (Fig. 8.2). For both these yield declines, the World Agricultural Price (WAP) for Crops is higher than the Base case, and this increases particularly in the second quarter of the next century (Fig. 8.3). The World Agricultural Price for Crops under the 20% yield decline shows signs of declining near the year 2050, while the decline under the 40% shows just the opposite. With more people to feed under declining crop yields, land will be increasingly shifted from forests (WFORST is world land area in forests compared to agriculture) as shown in Figure 8.4, more so if greater yield declines occur (i.e., 40 versus 20%).

World Cumulative Starvation Deaths (since 1990, in millions) in the Social Sector will more than double in the future in the declining yield simulation (Fig. 8.2), but there is no apparent effect on the World Population Growth Rate (100 million per year). Other variables examined, which showed no apparent difference, include World Calories Per Capita, World Physical Quality of Life, and World Gross Domestic Product. These are only hints at the kinds of analyses that can be done at the global scale. What is needed is a thorough evaluation of this and other models, examining not only total global conditions, but at national and regional levels.

Another model that could conceivably be used, on a national rather than a global scale, is Threshold-21.[22] Changes in crop production for example, from increased ground-level UV-B, ambient carbon dioxide and sulfur and nitrogen oxides, can be simulated and

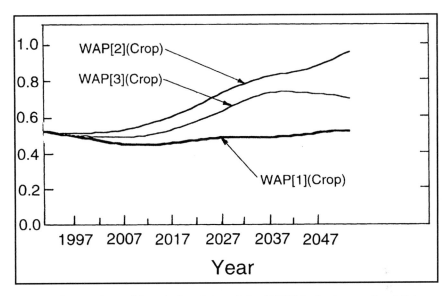

Fig. 8.3. Changes in World Agricultural Crop Price (WAP) in the International Futures Simulation (Professional Edition, Version 3.06), from year 1992 through year 2050, [1] no change in global crop yield (Base case); [2] gradual decline from 0.0% in 1992, to -40%; [3] and to -20% in global agricultural crop yield at year 2050. This is one of the connections of the Environment Sector with the Economics Sector in the IFS model.

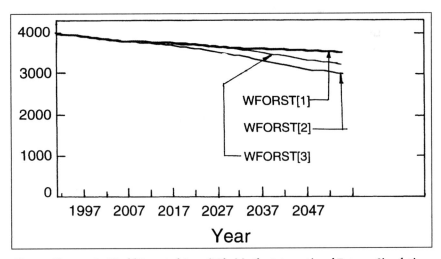

Fig. 8.4. Changes in World Forested Area (M ha) in the International Futures Simulation (Professional Edition, Version 3.06), from year 1992 through year 2050, with [1] no change in global crop yield (Base case); [2] gradual decline from 0.0 % in 1992, to -40%; [3] and to -20% in global agricultural crop yield at year 2050.

the resulting impacts seen in other economic sectors at a national level, including investments put back into social services, energy supply systems, the production of goods for trade, consumption and exports, military resources and quality of life.

The status of what is presently known regarding the UV-B climate across the world was discussed in chapter 1. Sparse evidence from observational data over the last several years on UV-B radiation in the Northern Hemisphere must be interpreted as showing increase at the ground level at least at some geographic locations. There is also observational evidence of a decrease in recent years at other locations. Similarly the evidence from computer modeling efforts shows conflicting results. One modeling study indicated an increase of 5 to 20% per decade in DNA-effective UV-B radiation at the latitudes of Europe from early springtime to summer respectively. Yet, another study indicated a decrease of 1 to 3% per decade.

On balance, when considering both the monitoring evidence and the modeling analyses, the most cautious scientific conclusion that can be drawn at the present time is that there may only be a slight, net increase of ground-level biologically effective UV-B radiation during the crop growing season at the northern latitudes. But, this situation could change in the future, especially if efforts to decrease emissions of air pollutants meet with success. The prudent course of action in the immediate future, would be to establish, increase and maintain a strong, long-term research and global monitoring effort for biologically effective UV-B radiation, especially during the spring and summer.

References

1. Cramer WP, Solomon AM. Climatic classification and future global redistribution of agricultural land. Climate Res 1993; 3:97-110.
2. Leemans R, Solomon AM. Modeling the potential change in yield and distribution of the earth's crops under a warmed climate. Climate Res 1993; 3:79-96.
3. Walker BH. Global change strategy options in the extensive agriculture regions of the world. Climatic Change 1994; 27:39-47.
4. Mannion AM. Agriculture in a warmer world. In: Agriculture and Environmental Change—Temporal and Spatial Dimensions. New York: John Wiley & Sons, 1995:335-341.
5. Weigel HJ, Kriebitzsch WU. Wirkungen von Klimaänderungen auf Agrar- und Forstökosysteme. In: Klimawirkungsforschung im Geschäftsbereich des BML. Schriftenreihe des Bundesministeriums

für Ernährung, Landwirtschaft und Forsten. Reihe A: Angewandte Wissenschaft. Heft 442. Münster, Germany: Landwirtschaftsverlag GmbH, 1995:43-59.
6. Wenkel K-O, Mirschel W. Systemanalyse und Modellierung—wichtige Elemente der agrarrelevanten Klimaforschung. In: Klimawirkungsforschung im Geschäftsbereich des BML. Schriftenreihe des Bundesministeriums für Ernährung, Landwirtschaft und Forsten. Reihe A: Angewandte Wissenschaft. Heft 442. Münster, Germany: Landwirtschaftsverlag GmbH, 1995:151-165.
7. Helms S, Mendelsohn R, Neumann J. The impact of climate change on agriculture. Climatic Change 1996; 33:1-6.
8. Fischer G, Frohberg K, Parry ML et al. The potential effects of climate change on world food production and security. In: Bazzaz F, Sombroek W, eds. Global Climate Change and Agricultural Production. Rome: FAO and New York: John Wiley & Sons, 1996:199-235.
9. Sombroek WG, Gommes R. The climate change—agriculture conundrum. In: Bazzaz F, Sombroek W, eds. Global Climate Change and Agricultural Production. Rome: FAO and New York: John Wiley & Sons, 1996:1-14.
10. Anonymous. Recommendations from Working Groups. In: Heidorn KC, Torrie B, eds. Proceedings of the International Conference on Ozone Depletion and Ultraviolet Radiation: Preparing for the Impacts, 27-29 April 1994. Victoria, BC, Canada: The Skies Above Foundation, 1995:215-225.
11. Anderson S. Global ecotoxicology: Management and science. In: Socolow R, Andrews C, Berkout F et al, eds. Industrial Ecology and Global Change. Cambridge, England: Cambridge University Press, 1994:261-275.
12. Hale B. UV impacts on food production. In: Heidorn KC, Torrie B, eds. Proceedings of the International Conference on Ozone Depletion and Ultraviolet Radiation: Preparing for the Impacts, 27-29 April 1994. Victoria, BC, Canada: The Skies Above Foundation, 1995:171-177.
13. Chatterjee K. CFCs and UV-B: Implications of Montreal Protocol. New Delhi, India: Environment Systems Branch, Development Alternatives, 1992.
14. Hoffman S. The World Almanac and Book of Facts 1992. New York: Pharos Books, 1991.
15. Park WI. The use of crop simulation models in the economic assessment of global environmental change. In: Biggs RH, Joyner MEB, eds. Stratospheric Ozone Depletion/UV-B Radiation in the Biosphere. Berlin: Springer-Verlag, 1994:311-313.
16. FAO. http://apps.fao.org/lim500/nph-wrap.pl?Production.Crops.Primary&Domain=SUA. 1997.
17. Caldwell MM, Teramura AH, Tevini M et al. Effects of increased solar ultraviolet radiation on terrestrial plants. Ambio 1995; 24(3):166-173.

18. Allen Jr LH. Plant effects and modeling responses to UV-B radiation. In: Biggs RH, Joyner MEB, eds. Stratospheric Ozone Depletion/UV-B Radiation in the Biosphere. Berlin: Springer-Verlag, 1994:303-310.
19. Fiscus EL, Booker FL. Is increased UV-B a threat to crop photosynthesis and productivity? Photosyn Res 1995; 43:81-92.
20. Parry M. Climate Change and World Agriculture. London: Earthscan Publications Ltd, 1990.
21. Hughes BB. International Futures—Choices in the Creation of a New World Order. Boulder, CO: Westview Press, 1996.
22. Hughes BB. Ifs—International Futures, Version 3.06, Professional Edition, CD-ROM for Windows 95. Boulder, CO: Westview Press 1997.
23. Millennium Institute. http://www.igc.apc.org/millennium/t21. 1997.

CHAPTER 9

Plant Population Genetics

James V. Groth

Introduction

Population genetics is the study of genes in populations. This field, which is largely mathematical and intuitive, was developed first from the simple mathematical descriptions of deterministic Mendelian traits. Although it overlaps, it is not synonymous with *quantitative genetics,* which deals with genes that act additively, so that the identity of individual loci is downplayed.[1] A third field that is becoming increasingly useful in describing agricultural systems is *ecological genetics.* Hartl and Clark[1] include ecological genetics in their broad definition of population genetics. Nevertheless, ecological genetics has a distinct history in that it is empirically based, much as agricultural science is. Ecological genetics deals with the role of genetic variation in ecological systems, and has been developed more inductively from genetic issues in ecology.

The aforementioned fields of science provide a framework from which to develop explanations, expectations, or hypotheses about the potential influences of increases in UV-B that can be expected in the future, if the stratospheric ozone layer continues to be depleted across all latitudes. However, the effects of elevated UV-B levels can only be considered at this time as an added stress factor on crops. In the context of the current status of knowledge, firm estimates are not possible of the direct and single effects of UV-B on crop populations or how UV-B will interact with other stress factors in the long range or on a large scale. Only a few general patterns of what might be possible can be described. Hence, much of the discussion will center on the degree of robustness (their ability to economically

---

Elevated Ultraviolet (UV)-B Radiation and Agriculture, by Sagar V. Krupa, Ronald N. Kickert and Hans-Jürgen Jäger.
© 1998 Springer-Verlag and Landes Bioscience.

produce under a variety of conditions) of high input/output agricultural systems to additional stress in general. One other caveat about predicting the direct and single effects of UV-B: it may prove to be an academic exercise, since increased UV-B is unlikely to occur unaccompanied by alterations in other anthropogenic stresses, including changes in other meteorological variables, all of which comprise the climate.

The Nature of Intense Cropping Systems

High input/output cropping systems of the developed world have evolved as a result of the societal desire to maximize productivity on available land. Short range goals and considerations have almost always determined how these cropping systems develop. Short range considerations are also usually paramount in the economic forces that maintain these systems, from the need for growers to maximize immediate profits and stabilize cash flow, to the market forces that govern prices and uses of the commodities.

In general, the more intensive cropping systems are those with minimal genetic diversity in a small number of species that can be rotated to reduce losses to pathogens and pests and the depletion of soil nutrients. Intensive systems are at the end of a process of gradually narrowing the genetic base as crop species are domesticated (Fig. 9.1)In more developed agricultural systems, in the final stage of this process, monoculture seems to be the rule.[3] If additional pest and disease reduction is needed, it is achieved through genetic resistance or application of pesticides, superimposed on the existing genetic uniformity. Attempts to introduce intra-field diversity into crops, for any reason, have not been widely adopted. On the contrary, some crops that were originally developed and grown as diverse, open-pollinated populations (e.g., corn, *Zea mays* L.) have been turned into monocultures of hybrids, because of the increased, more predictable yield potential and the greater suitability for mechanization. Table 9.1 shows the state of uniformity in the major crops. These data were compiled in part in response to the U.S. northern corn leaf blight epidemic that resulted from extreme genetic uniformity of the 1970 corn crop. The problem of crop uniformity has not improved since then. Moreover, in many crops, the extremes of the range in which the crop can be grown have even less diversity than the more suitable centers of the crop range, because: (1) less effort has been expended in breeding the crop for the marginal habitats of

Plant Population Genetics

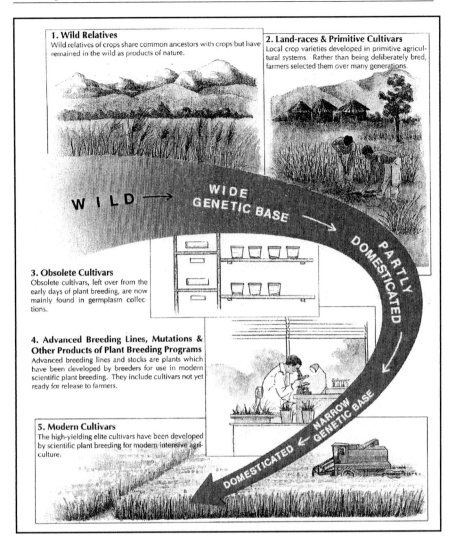

Fig. 9.1. Illustrating the steady decline in genetic diversity as wild plant species are increasingly domesticated.[2] See color insert for color representation.

the range extremities (often the crop has not been grown as long there) or (2) there has been less suitable germplasm to begin with for marginal habitats, and so cultivars tend to have more similar ancestry at the edges of the range than in the center. Both corn and soybeans (*Glycine max* (L.) Merr.) show these tendencies in the corn belt states of the U.S.

Table 9.1. *The degree of genetic uniformity in commercial cultivars of major U.S. crops in 1969*[4]

Crop	Acreage (millions)	Value (millions of dollars)	Total number of varieties	Number of major varieties	Acreage (percent)
Bean, dry	1.40	143	25	2	60
Bean, snap	0.30	99	70	3	76
Cotton	11.20	1200	50	3	53
Corn[a]	66.30	5200	197[b]	6	71
Millet	2.00	---	---	3	100
Peanut	1.40	312	15	9	95
Peas	0.40	80	50	2	96
Potato	1.40	616	82	4	72
Rice	1.80	449	14	4	65
Sorghum	16.80	795	?	?	?
Soybean	42.40	2500	62	6	56
Sugar beets	1.40	367	16	2	42
Sweet potato	0.13	63	48	1	69
Wheat	44.30	1800	269	9	50

[a]Corn includes seeds, forage and silage. [b]Released public inbreds only.

Traditional breeding continues to favor elite germplasm with a narrow genetic base. The potential for genetic engineering to increase diversity has not been met, at present. In fact, some technological tools have been used to increase uniformity of crops. For example, asparagus (*Asparagus officinalis* L.) was traditionally grown from seed, and since asparagus is an outcrossing species, all cultivars retained genetic diversity, at some cost in yield or quality. Advances in clonal propagation via tissue culture have allowed growers to switch to uniform stands of high quality genotypes.

Population Genetics of Crops

Because of their genetic uniformity, coupled with extreme homozygosity (or identical levels of heterozygosity in individual plants for hybrids) in many crops, many of the genetic concepts for natural populations cannot be applied to the crop population. This includes such central premises as *random mating, fine scale polymorphism, and Hardy-Weinberg equilibrium*. Seeds that give rise to crop stands have been produced under controlled conditions in order to ensure uniformity, and random mating of plants (either artificially or through open pollination) is not permitted. Ideally, new genotypes and genetic variation that results from random mating or even spurious outcrossing are absent from such stands. Within the stand, there is no genotypic or phenotypic polymorphism (the occurrence of two or more forms at frequencies of over 1% each). Likewise, with no random mating, there can be no meaningful Hardy-Weinberg equilibrium (the steady state frequencies of the homozygotes and heterozygotes for any locus that is a function only of the frequencies of the individual genes). In self-pollinated crops there are no heterozygotes, whereas, in hybrids, many loci are 100% heterozygous.

If genetic variation is the raw material for selection, then there is very little to work with in most crops. Rather than allowing forces of natural selection to mold and change a crop population, humans have taken over this role, using a totally different set of tools and methods than would be found in natural populations. Population genetics attempts to model the multi-dimensional changes that can take place in changing populations of freely breeding plants. Within population genetics, only a limited body of quantitative genetic theory has been developed to model plant improvement. Complex theories in population genetics are not proven. Often they revolve around mathematical arguments (derived originally from Mendelian genetic principles) that have only a minimal biological basis. Parameters are handled simply, and reflect general principles without allowance for exceptions. Models are often deterministic rather than stochastic. In summary, even those population genetics concepts and models that can be applied to crop populations often will have little predictive value. Some of these concepts are addressed below.

Robustness of Populations and Communities with Genetic Diversity

Populations of single or multi-species communities are said to be rich when they are composed of many genotypes or species, respectively. Richness is often equated with robustness or resiliency, which is the ability of the population or community to rapidly adapt and maintain itself in the face of changing environments. In its simplest form, this phenomenon involves having a range of genotypes or species on hand (or for sexually reproducing organisms, generating them quickly) that can assume more dominant roles, if they find themselves to be most fit in a new environment. The details of this process can be argued (cf. Maynard Smith[5]), but in general, rich or diverse populations are more flexible in adapting quickly to changes in their environment than are clonal or monotypic populations.

But robustness is lost in crop populations, because the tacit assumption in developing crops for commercial production is that natural selection is not influential or desirable—humans are in control of the crop. Scientists are aware of the fallacy of this view—it represents an ideal that is only approximated. Growth of the crop and yield in a specific environment are dependent on a type of adaptation of a genotype to the environment. Through observing superior performance in yield trials, plant breeders practice artificial selection. Resultant genotypes are sometimes robust, and find use in a wide spectrum of environments. In many crops, a few inbred lines or cultivars have been widely used alone or in hybrids, because they perform well in large areas (Table 9.1). This phenomenon may be fostered by uniformity in cultural practices. Often, however, genotypes that are superior in an environment are narrowly adapted only to environments and cultural conditions similar to those in which they were selected.[6] Such germplasm can be said to lack robustness.

Elite germplasm (that which is capable of responding to high inputs of fertilizer and, in favorable environments, realize very high yields) from highly specific conditions of very intensive agriculture, tend to lack the kind of robustness that characterizes widely adapted land races, or what is found in multiple genotype or multispecies cropping systems of less intensive agriculture. This is due partly to the emphasis in intensive agriculture on uniformity for maximum response to inputs with minimum negative competition between plants and on the great gains in efficiency of harvest that single passes with farm machines provide. During the evaluation and selection processes of plant breeding, elite lines are only viewed under near ideal conditions, that closely match the conditions under which the crop is to be grown. While the intent is not to exclude more resilient materials, this often happens (Fig. 9.2). The result is that elite materials usually do not perform well under more marginal conditions. In other words, these lines are adapted to a very narrow set of environmental conditions. While most of them are untried in environments where additional stresses due to climate change are present, their performance in such environments is not likely to be good.

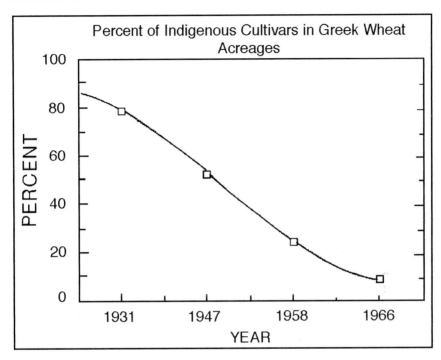

Fig. 9.2. An example of the steady displacement of old land race germplasm, which is characterized by good local adaptation, higher genetic diversity and modest yield potential, by modern elite germplasm, characterized as less adapted, more genetically uniform and capable of responding to high inputs with high yields, in wheat (*Triticum aestivum* L.) cultivars in Greece.[7]

Buddenhagen[6] illustrated this in the closely analogous case in agriculture of marginal environments of the developing world. These environments include poor, often saline soils, poor nutrient regimes, additional disease and insect pressures and uneven moisture patterns. He concluded that crops must be evaluated under such conditions in order to efficiently obtain the best performers.

Most elite crop populations have little or no genetic variation to permit any change or adaptation as the environment changes. They are completely dependent on plant breeders to provide this adaptation. Indicative of this is the observation that many elite lines of crops have lost the capacity to survive on their own and fail to appear during the next season as volunteer plants.

Plant Breeding as a Mimic to Natural Processes of Adaptation

If crop populations have lost the capacity to adapt on their own, they have an alternative means to change—the ability of plant breeders to adjust the crop to the environment. Methods of improving plants differ greatly for different crops. The basis for the differences in breeding crops is a function of biological differences, especially the tolerance of a species to extreme inbreeding and the ease of obtaining mass exchange of pollen, and economic differences, including the way in which the crop is to be harvested and used. In addition, methods are chosen that maximize progress with minimal cost. Often cost and progress (measured as rate or magnitude of change in important traits) are competing forces, and compromise strategies are adapted. Two examples can be used in this context. Single gene, race-specific resistance to rust diseases in cereals is often chosen over more durable polygenic partial resistance, because single-gene resistance is much easier to incorporate into adapted backgrounds via backcrossing. This is a common short-term solution to a disease problem, however, it is considered by many to be more effective even in the long-term, because new genes for resistance can be obtained and incorporated in series relatively easily for many rust diseases. The other example is the general issue of the level of diversity to be included by breeders in the germplasm to be used as raw material in crossing blocks or open pollinated populations for new elite cultivar development. Depending in part on the specific breeding methods that are used, higher levels of diversity (inclusion of wild or ancestral species, for example) tend to slow progress toward acceptable cultivars, even though the distant germplasm may bring in important traits that have been lacking in the crop. Often distant sources are included only when they are known to contribute such a trait and the trait is desperately needed.

Plant breeders can be quite resourceful in adapting breeding schemes to the immediate need to maintain productivity, given that they have sufficient warning of change. Some of the methods they can use and their analogs in natural systems are discussed below.

Seeking Sources of Germplasm that Can Tolerate the New Environment

In most diverse natural environments, some degree of tolerance to an environmental change may exist already, or may exist in part in several genotypes. Strong selection pressure for tolerance combined with genetic recombination can result in rapid dominance of new and tolerant genotypes. Things are not likely to happen so rapidly in crop species. Since there is restricted variability in the elite germplasm of a crop, the likelihood that suitably tolerant forms exist is diminished. If tolerant cultivars can be found, simple replacement of current cultivars with the tolerant form may be possible, which will provide a rapid and effective solution. This is indeed a possibility for soybeans, where the cultivar Williams is reported to tolerate elevated UV-B better than the cultivar Essex, and is of good agronomic quality.[8] The demonstration of this variation in a rather narrow genetically based crop (see Table 9.1) is rather surprising.

More likely, the location-specific nature of much elite germplasm will require that a tolerance trait like that in Williams be transferred into one or more other elite backgrounds. This can be accomplished through backcrossing in a year or two if inheritance of tolerance is simple, or through other breeding methods if heritability is high. Otherwise, it will require more time (perhaps 10 years) and resources. Presently, nothing is known about the mode of inheritance or heritability of UV-B tolerance, nor is this UV-B tolerance considered an important agronomic trait by plant breeders.[8]

If tolerance is absent in elite germplasm, it may need to be brought in from elsewhere. In natural populations, this happens when outside germplasm migrates into an area via some intermittent agent of dispersal. In crop plants, humans seek out tolerance and incorporate appropriate sources into breeding programs. There are two problems with this. First, maximum diversity in a crop (or closely related) species is found in the centers of origin or in the secondary centers of diversity of the crop.[9] However, even if there is great diversity for most traits in such places, there is no reason to expect diversity for tolerance to environmental factors that have not exerted selection pressure on the species there. Other areas, where the environmental factor is present, require searching. Experience with disease and insect resistance certainly supports this.[10] Secondly, exotic germplasm will probably be genetically distant from desired elite germplasm, being often from a wild progenitor species or even from an unrelated species. This will complicate transfer of the tolerance (Fig. 9.3). If the inheritance of tolerance is complex, it may require great effort to transfer from distant source backgrounds using conventional breeding methods. At the worst, a compromise may have

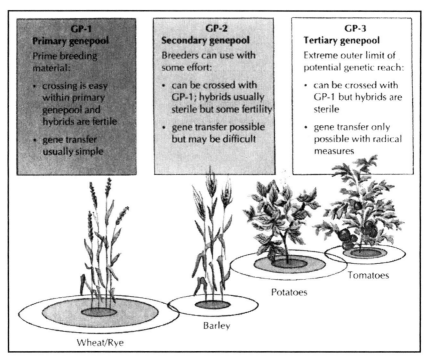

Fig. 9.3. The three categories of germplasm available to plant breeders, classified according to the degree of relatedness to and ease of introgressing genes into the crop. GP-2 and GP-3 levels are difficult to use with more elite crop cultivars of intensive agriculture.[2]

to be reached in which only part of the tolerance is transferred in order that the quality of the crop be retained. The likelihood of transferring all tolerance into a desired background will be directly proportional to: (1) the size of the progeny populations that can be screened and (2) the heritability of tolerance, inversely proportional to: (a) the genetic distance between source background and elite background, (b) the number and distribution of genes within the genotype determining the tolerance and (c) the specificity of agronomic or horticultural characters of a crop dictated by the need for high yields or narrow quality standards. Developments in genetic engineering may increase speed and efficiency of transfer of traits between species. At the present time, this promise has not been realized.

Genotype by Environment Interactions

The issue of whether elite germplasm is necessarily likely to also be narrowly adapted germplasm, has been a key question. Broad adaptability has been a recent goal with elite germplasm, driven and inspired, perhaps, by the successes of the large international breeding programs that have resulted in great yield increases in a small germplasm base, for wheat (*Triticum aestivum* L.) and rice (*Oryza sativa* L.) in much of the world, where intensive agriculture can be practiced.[6] The idea is to produce breeding lines that can be taken around the world and modified by local plant breeders to produce suitable, high yielding cultivars for the specific environments of each country or region. But is this germplasm really broadly adapted, or is this simply a successful, but dangerous short-term solution?

One can ask whether the agricultural methods are not being modified (in the sense of being made more uniform everywhere) as much or more than the germplasm is being modified. If additional stresses materialize in a relatively short time, will such worldwide strategies be able to respond rapidly or effectively? Natural plant populations sometimes respond to change only after their size is greatly reduced. This is not an option with crop plants. The nucleus of the problem is that we need to maintain at least moderate, and preferably high productivity during the time that we are trying to modify the crop. This may be impossible with some crops, unless we can broaden our selection goals to include more than one "ideal" genotype. If we try to develop stress-tolerant crops that are identical in all other respects to the original crops, and are designed to be grown in identical cropping systems as before, prospects of maintaining high productivity during this process and regaining previous levels of very high productivity, seem poor. Alternatively, if the entire system of cropping, including the nature of the elite germplasm, can be made more flexible, so that more than one target ideal genotype is available for breeding efforts, it would increase the chance of regaining high productivity in a relatively short time.

The Predicted Effects of UV-B on Crop Populations

Many effects of elevated levels of UV-B on plants have been documented, at least under artificial conditions.[8,11] The levels of UV-B used in experiments are apparently consistent with expectations based on predicted levels of stratospheric ozone depletion. This is useful as long as the effects are compared with present day levels of

UV-B, rather than with control plants that receive **no** UV-B. The nature and severity of effects vary from species to species. Moreover, other environmental parameters can influence the effects of UV-B in various ways, both positively and negatively.[8,12] Because the effects cannot be generalized, here discussion is directed to the negative possibilities, and UV-B is treated as an additional stress factor on crops. It is also assumed that in the short-term and, perhaps less so, in the long-term, the nature of intense crop production will not change significantly. This is based on the expectation that the influence of UV-B alone on crop productivity will not be catastrophic, nor will the effects of other anticipated global climate changes.

Long-Term Expectations

If levels of UV-B increase gradually and smoothly over a long time period of, say, 50 years or more, prospects for maintaining good agricultural productivity in the face of this rise will remain high. It is safe to say that other characters that contribute to the non-sustainability of high intensity cropping systems will be much more limiting than will climate change. Some examples are environmental pollution by fertilizers and pesticides, soil nutrient depletion and erosion, and ground water depletion. Solutions to these problems will be especially difficult if we are to also maintain the high levels of crop production per hectare that we have enjoyed, up until now. As a direct cause of loss in productivity, it does not seem likely that UV-B is going to rank with other anthropogenic changes, which means that gradual increases in UV-B will be more easily accommodated in agroecosystems. This assumes that peak levels of UV-B during episodic events will not be extremely high. In the long run, problems with our crops stemming from a narrow intra- and interspecific base can be overcome, if we so choose. Moreover, selection for tolerance to UV-B, given that there is adequate variation in the elite germplasm, may occur gradually and unwittingly, as poor performing materials (including those that perform poorly because of sensitivity to UV-B) are eliminated from breeding programs. If, on the other hand, sufficient variation does not exist in locally elite materials of a crop, more conscious efforts to seek out sources of tolerance will be necessary, perhaps from areas where UV-B levels are high and where the crop is grown, so that natural selection (for related native plants see Caldwell et al,[13] Bornman and Teramura,[14] Rozema et al[15]) or artificial selection for crop plants, will have already resulted in elevated tolerance levels. Particularly sensitive crop species may be eliminated from areas where they are presently grown, because of various global changes, including UV-B. Present day cropping systems have been shaped by existing environmental forces. For instance, most processing sweet corn is grown in the northern states of the U.S., because the number of disease and insect problems in the southern states often reduced the quality of the product so much that the advantages of the southern climate were more than offset. In an environment modified by humans, trial and error will continue to be used to adjust where crops can be grown, and if necessary, new, more stress-tolerant crop species may replace old ones in certain districts. Sometimes forced change can be a blessing, and our response to climate change may divert more research resources toward solutions that also raise system-wide questions about how we obtain high yields.

Short-Term Expectations

In the event that mean or episodic levels of UV-B become elevated to the point of causing large losses in productivity in a relatively short time (say 20-30 years), we will probably have to suffer those losses for some time; agriculture will be hard pressed to overcome the problems in the short run. Two kinds of uniformity can be implicated and are described in the following sections.

Interspecific uniformity

Agroecosystems differ from natural ecosystems in that the number of species that comprise agroecosystems is much smaller, and the number of species that occupies the "community" is usually one (ideally). Much of the resiliency of natural communities resides in their ability to respond to change by adjusting their species composition. This option has been totally removed from agroecosystems, except in the special case of complete replacement of crop species during rotation, or in the rare instance when a new crop is developed that displaces one on which dependence was previously placed.

Humans are dependent for primary food production on a few key crop species. The choices that modern growers have are severely limited; more so in high latitudes, where short growing seasons and lack of cumulative heat units prevent some of the important species (corn and soybeans, for instance) from being grown. If a particular commodity is especially susceptible to UV-B, options for replacing it with a different, existing commodity are few in the more marginal agricultural areas. Another possibility is to find new commodities that are more tolerant of UV-B. New species are constantly being examined, and several are in various stages of development. Still, the number of new species that have emerged in modern times as major production species worldwide is very small. Complexities abound in this process. Markets must be developed or explored, optimum conditions for growing the crop must be determined by extensive field experimentation, and machines must be engineered to harvest and handle the crop, among other things. Farmers and lending institutions cannot afford to risk their livelihood on unproven crops. They must be convinced that the crop will be profitable before they will accept it as an alternative to existing crops. It does not appear to us that there is any way to appreciably speed up the process of getting new crops on line. Only after a crisis occurs do systems like this begin to change, and then it is usually too late.

Intraspecific uniformity

The other half of the monoculture problem is that most of our major crop species lack genetic diversity. As stated previously, if at all possible, we have opted to grow our crops as single-genotype stands. This is true for some of the most important crops: wheat, rice, barley (*Hordeum vulgare* L.), corn and soybeans. Often different cultivars, even those developed and grown in distant parts of the earth, share significant amounts of their pedigrees. Barrett[3] discusses some of the reasons for this, which usually come down to convenience to all people concerned and efficiency of the breeding strategies. The consequences of this—rapidly changing our crops to accommodate changes in their environments—re several: it is less likely in genetically narrowed crops that adequate levels of UV-B tolerance will be found in elite, desirable backgrounds. Progress in incorporating tolerance will be slowed if: (1) tolerance sources are genetically distant from the elite germplasm and, (2) tolerance is genetically complex. Such tolerance can certainly be incorporated into elite backgrounds in time, with the process being speeded if the size of the breeding program is increased.

There is room for optimism about being able to deal with UV-B or other stress factors that may be on the horizon. New molecular developments in plant breeding may allow more rapid progress in moving traits from wild relatives to crop species. In a sense, we are witnessing a race between our ability to anticipate and deal with impending "new" stresses on our crops and the arrival of those stress factors.

References
1. Hartl DL, Clark AG. Principles of Population Genetics, 2nd ed. Sunderland, MA: Sinauer, 1989.
2. Hoyt E. Conserving the Wild Relatives of Crops. Rome: IBPGR Headquarters, C/O FAO of the United Nations, 1988.
3. Barrett JA. The evolutionary consequences of monoculture. In: Bishop JA, Cook LM, eds. Genetic Consequences of Man-made Change. London: Academic Press, 1981:209-248.
4. National Academy of Sciences. Genetic Vulnerability of Major Crops. Washington, DC, National Academy of Sciences, 1972.
5. Maynard Smith J. What use is sex? Theor Pop Biol 1971; 30:319-335.
6. Buddenhagen IW. Breeding strategies for stress and disease resistance in developing countries. Annu Rev Phytopathol 1983; 21:385-409.
7. Hawkes JG. The Diversity of Crop Plants. Cambridge, MA: Harvard University Press, 1983.
8. Tevini M, Teramura AH. UV-B effects on terrestrial plants. Photochem Photobiol 1989; 50:479-487.
9. Leppik EE. Gene centers of plants as sources of disease resistance. Annu Rev Phytopathol 1970; 8:323-344.
10. Wahl I, Anikster Y, Manisterski J et al. Evolution at the center of origin. In: Bushnell WR, Roelfs AP, eds. The Cereal Rusts, Vol. I: Origins, Specificity, Structure and Physiology. Orlando, FL: Academic Press, 1984:39-77.
11. Krupa SV, Kickert RN. The Greenhouse Effect: Impacts of ultraviolet-B (UV-B) radiation, carbon dioxide (CO_2), and ozone (O_3) on vegetation. Environ Pollut 1989; 61(4):263-393.
12. Krupa SV, Kickert RN. The Effects of Elevated Ultraviolet (UV)-B Radiation on Agricultural Production. A peer reviewed critical assessment report submitted to the Formal Commission on "Protecting the Earth's Atmosphere" of the German Parliament, Bonn, Federal Republic of Germany, 1993.
13. Caldwell MM, Robberecht R, Nowak RS et al. Differential photosynthetic inhibition by ultraviolet radiation in species from the arctic-alpine life zone. Arctic Alpine Res 1982; 14:195-202.
14. Bornmann JF, Teramura AH. Effects of ultraviolet-B radiation on terrestrial plants. In: Young AR, Moan J, Björn LO et al, eds. Environmental UV Photobiology. New York: Plenum Press, 1993:427-471.
15. Rozema J, Gieskes WWC, van der Geijn SC et al, eds. Special Issue: UV-B and Biosphere. Plant Ecol 1997; 128:1-313.

CHAPTER 10

Genetic Engineering of Crops

Introduction

Genetic engineering is the use of recombinant DNA technology to make heritable changes in the DNA of an organism. The development of requisite tools and technologies, combined with an increased understanding of molecular genetics and biology has made it possible to create genetically transformed plants that carry introduced uniform genetic traits. More than 40 species of transgenic food and fiber crops have been created and approximately 600 field tests have been or are being completed in more than 20 countries around the world.[1] The use of transgenic plants in agriculture has great potential for increasing agricultural productivity, reducing crop losses and improving food quality and safety.[2] Use of transgenic plants for insect, disease and herbicide resistance[3,4] control were predicted earlier and some such plants are now commercially available in the U.S.[5-9]

The preceding summary demonstrates the feasibility of applying modern biotechnology techniques to create genetically transformed plants capable of reducing potential negative effects of increased UV-B radiation on plant tissues. Unfortunately, the actual effects of increased solar UV-B on plants under field or ambient conditions is not adequately known at this time. However, Caldwell et al[10] and Krupa and Kickert[11] have discussed the potentially deleterious consequences for crop productivity. Increased UV-B irradiation has been shown to result in yield reductions in a specific cultivar of soybeans (*Glycine max* (L.) Merr) in one study.[12] In barley (*Hordeum vulgare* L.), on the other hand, a naturally occurring, UV-induced mutation resulted in durable resistance to the powdery mildew fungus *Erysiphe graminis* f. sp. *hordei*. In fact, this mutation has been bred into many of the spring barley lines currently in use throughout Europe.[13]

Elevated Ultraviolet (UV)-B Radiation and Agriculture, by Sagar V. Krupa, Ronald N. Kickert and Hans-Jürgen Jäger.
© 1998 Springer-Verlag and Landes Bioscience.

Whether or not increased UV-B irradiation will occur at levels significant enough to damage crop plants in the near future at all latitudes has not been conclusively demonstrated (refer to chapter 1). Thus, although the following discussion provides information on what genetic engineering could do to offset yield losses caused solely by increased UV-B, it may be an academic point. In addition, if conditions are present which result in increased UV-B irradiation, other growth regulating variables most likely will also have been altered and may have as great an effect on crop productivity as the increase in UV-B.

Targeting Potential UV-B Damage in Plants

The most efficient means by which genetic engineering may offset yield reductions resulting from UV-B irradiation is to target a specific area for "protection" or repair (i.e., what needs to be protected or repaired, be it DNA, proteins, or structural components of the host and the tissue or organs in which the target resides, needs to be clearly identified). Runeckles and Krupa[14] identified five primary levels of effects of increased UV-B irradiation on plant growth. These five categories include: (a) biochemical (DNA, RNA, proteins, peptides and phytohormones); (b) structural (membranes and chloroplasts); (c) functional/physiological (photosynthesis and stomatal function); (d) accumulative (growth, morphogenesis, yield, flowering and reproduction); and (e) ecological/environmental.

Of these five levels, the areas most affected, and certainly the areas resulting in the largest reductions in yield, following increased UV-B irradiation are the biochemical and structural components. Electromagnetic radiation in the UV-B (280-320 nm) can induce genetic damage in cells with both lethal and mutagenic consequences.[15] In addition, UV-B can transform eucaryotic cells to a heritable altered state. Among the types of mutations inducible by UV-B are: cyclobutane pyrimidine dimers,[16-18] single strand breaks,[19] DNA-protein crosslinks,[20] and a 5-hydroxy-methylcytosine and other DNA photoproducts.[21,22]

According to Bornman[23] and Strid et al,[15] one of the major sites affected by UV-B is the chloroplast, with UV-B induced changes to the biochemical composition of the chloroplast and the impairment of photosynthetic function. Jordan et al[24] demonstrated that maximum ribulose-1,5-bisphosphate carboxylase (Rubisco) activity declined by 71% after three days of increased UV-B exposure in pea plants (*Pisum sativum* L., cv. Greenfeast). Additionally, mRNA transcript levels of Rubisco were reduced by more than 60% following exposure to UV-B radiation.[24]

Thus, two major goals for engineering plants with decreased sensitivity to increased levels of UV-B are protection of the DNA complement of plants and protection of the photosynthetic apparatus.

Methods of Crop Protection

There are two basic strategies that can be used for the protection of crop plants against increased UV-B radiation. The first of these is avoidance or dilution. To counteract the losses incurred from the effect of increased UV-B on the photosynthetic apparatus, plant morphology or structure could be modified. Such modifications could include changing the plant architecture to provide more leaves per plant, or add/increase the number of trichomes on the leaf surface. Additionally, elongation and consequently, thickness of the leaf epidermal cells could be increased. Each of these modifications would attempt to decrease the effect of UV-B by either blocking more UV-B or by providing more photosynthetic area to offset losses incurred. This strategy could be implemented using genetic engineering, traditional breeding methods, or a combination of both.

The second strategy would be to augment plant DNA by the addition of DNA from a source other than the parental germplasm, using recombinant DNA techniques. The transferred DNA would confer protection to host DNA by preventing or repairing possible damage caused by UV-B irradiation. This strategy will be the focus for the remainder of this discussion.

Genes Involved in the Repair of DNA Damaged by UV-B Irradiation

Several genes have been identified in bacterial systems which are involved in the repair of UV-B damage, as deduced from survival curve measurements.[19,25-27] In general, most of these repair genes work by excising and then repairing induced dimers.

Okaichi et al[28] demonstrated repair of DNA of the cellular slime mold, *Dictyostelium discoideum*, damaged by UV-B irradiation. Induced pyrimidine dimers were removed from DNA molecules by excision repair. Keszenman-Pereyra[29] showed that repair of UV-damaged DNA in the yeast, *Saccharomyces cerevisiae*, is carried out primarily by excision repair processes. Additionally, a gene product was identified which may be essential for the processes involved in the repair of chromosomal DNA.

Sutherland et al[17] determined the capacity of soybean seedlings to repair DNA damage by UV doses that do not produce apparent injury in the plants. The repair involved the removal of cyclobutane pyrimidine dimers by both excision and photoreactivation. Photoreactivation was detected in seedlings at all initial dimer levels. Although excision was not observed at the lowest dimer frequencies, at higher initial dimer levels, it was quite effective in the dimer removal. The rates of repair in soybean were substantially higher than in alfalfa (*Medicago sativa* L.) seedlings at the same DNA damage levels.

According to Jordan et al,[30] the response to increased levels of UV-B radiation is dependent upon the developmental stage of the tissue and involves complex changes in gene expression. Nevertheless, identification of genes responsible for the repair of UV-induced DNA damage needs to be the first step in the transfer of those genes into plant species, thus conferring increased resistance to elevated UV-B irradiation. Another strategy is to increase the concentrations of UV-protectant pigments.

Plant Transformation

Several plant transformation systems are currently in use. These systems are divided into two groups; direct DNA transfer systems and the *Agrobacterium tumefaciens*-mediated transfer systems.

Direct DNA Transfer Systems

One of the primary reasons for interest in direct DNA transfer technology is that only a few monocotyledonous species are susceptible to *A. tumefaciens*, therefore the *Agrobacterium* transformation system cannot be applied to most monocots. The effective transformation of a plant by direct DNA transfer, using either biolistic, electroporation, or protoplast fusion techniques, requires that the experimental tissue be forced through a period of cellular dedifferentiation and subsequent plant regeneration from calli. This period of dedifferentiation is not a requisite for all plant species as evidenced by the transformation of soybeans using the biolistic approach on apical meristems. The need for plant regeneration has limited the successful application of direct transformation techniques of most monocotyledonous species, because regeneration of these species has proven to be much more difficult than the regeneration of dicotyledonous species. However, several examples of regenerated monocots including corn (*Zea mays* L.), rice (*Oryza sativa* L.), wheat (*Triticum aestivum* L.) and oats (*Avena sativa* L.) have been reported. Though the regenerated monocots have usually been limited to a specific genotype within a given species, the fact that plants have been regenerated, and in some cases transformed prior to regeneration, indicates that the problem of cellular nontotipotency in monocots may be overcome.

Agrobacterium-Mediated Transfer Systems

Agrobacterium tumefaciens, a soil-borne bacterium has the ability to induce tumors on all dicotyledonous plants tested to date. Tumor formation results from the transfer, insertion, and expression of a segment of DNA from the tumor inducing, Ti plasmid carried by *A. tumefaciens*, into the plant cell genome. The segment of the Ti plasmid DNA which is transferred from the bacterium and becomes stably integrated into the plant genome is known as T-DNA. The Ti plasmid can be modified so that genes of interest, such as the aforementioned DNA repair genes, can be transferred to plants.

The introduction of foreign DNA into plants using the *A. tumefaciens* transfer system follows several general steps. The Ti plasmid is disarmed by removing genes for tumor induction, then the T-DNA region is subcloned into a small conventional plasmid vector. The gene of interest is then inserted into the T-region and the modified plasmid transferred into *Escherichia coli*. The modified plasmid is then transferred to *A. tumefaciens*. The transfer of the T-DNA is brought about by a tri-parental mating involving the modified plasmid containing *E. coli*, an *E. coli* carrying a transfer helper plasmid and the recipient *Agrobacterium*. Once the modified T-region has been introduced into the *Agrobacterium*, in vivo homologous recombination inserts the T-region into a resident non-recombinant Ti plasmid. The actual current use of *Agrobacterium* transfer systems employs the use of vectors to insert the DNA of interest into the T-DNA of the bacterium, however, the resultant modified bacteria are much the same as that derived from traditional methods. The virulence genes of the resident Ti plasmid are responsible for the transfer of the recombinant T-DNA into the plant genome. This transfection of the host plant is accomplished by the cocultivation of the host with the modified *Agrobacterium*. Following this culture period, whole plants are regenerated and tested to determine the success of the transformation process.

Gene Expression in Transformed Plants

Though foreign genes can be inserted in target plants, their expression and/or function may differ from that which is expected. Gene expression in transformed plants can be influenced by many factors. The position and number of integrations of the foreign sequence within the plant genome have been identified as the most important factors influencing gene expression.[31] In addition to position of insertion, the transcription of inserted genes may be dependent on other transcription factors, such as *cis*-acting elements, which may or may not be part of the recipient's genome. Not only quantitative, but also qualitative changes in gene expression may occur.[32] These qualitative changes may result in altered organ-specific expression patterns. Qualitative effects may be due to regulatory sequences around the insertion site of the foreign sequence.[31]

The Role of Molecular Biology

Perhaps the most important role of the new technology is not the actual engineering of crop species, but rather to aid traditional breeding programs in crop improvement. Decreases in crop yields due to increased UV-B irradiation would most likely result from a polygenic effect. Thus, several or many genes would be involved in the yield decline. The most effective way to work with polygenic traits is through traditional plant breeding. Many of our modern crop species have lost much of their genetic variation due to intensive breeding programs (refer to the previous chapter in this book). However, genetic diversity still resides within the native or land race species from which our modern cultivars have been derived. The problem faced by breeders in introducing variation into cultivar populations from wild relatives is that the genetic load of undesirable characteristics introduced at the same time requires years of backcrossing and selection to introduce only desired characteristics into a line which would result in a commercially acceptable variety.

Through the use of molecular biology, it is possible to identify genetic markers, i.e., segments of DNA associated with a specific phenotype. The use of genetic markers can greatly enhance traditional breeding programs. To obtain these markers, populations of a crop species and their relatives could be tested to ascertain their yield performance under specific cultural, including environmental conditions. Those that performed well would be genetically dissected in an attempt to find a polymorphic region of DNA associated with desired performance in the yield tests. If a DNA segment was found which was often associated with good performers, this segment could be used as a genetic marker in traditional breeding programs.

Breeders could make many crosses between current cultivars and individuals which did well in the cultural tests. Traditional selection would be applied to the resulting progeny. It is at this point that the use of molecular techniques can rapidly improve breeding populations. Once selections have been made, the selected individuals are genetically dissected and their DNA hybridized to that of the genetic marker. Those individuals that carry DNA sequences homologous to the marker sequences have a far greater chance of carrying the DNA sequences responsible for the better performance characteristics, under the test conditions, of the parent from which the marker was derived. The selected progeny that also carry the marker sequences would then be carried into a backcross program. Thus, the use of genetic markers allows more intensive selection in the progeny than would otherwise be employed, resulting in a large decrease in the backcross breeding population.

A Perspective of the Future

The tools and technologies for producing transgenic plants are currently available. However, the knowledge of possible field effects of increased UV-B irradiation on crop plants is deficient. Genetic engineering requires some specificity as to the type of genetic change requisite in order to achieve a specific goal. Therefore, it is imperative that we increase our understanding of the impact of increased UV-B on crop plants before attempting to engineer plants to compensate for the potential losses induced. If indeed, increased UV-B irradiation has a profound detrimental effect on crop plants, then the effects must be determined as soon as possible. The period between identification of the problem, and a genetically engineered response in crop plants could have a lag time as great as ten years.

Though it may be possible to engineer plants to withstand increased UV-B irradiation, the conditions that result in that increase undoubtedly will alter the patterns of other growth regulating variables that could have as great an effect on crop productivity as the increase in UV-B. Thus, the ability to engineer UV-B-resistant crop plants may, in the end, be nothing more than an academic question.

The real benefit of modern molecular techniques may be the ability to find genetic markers associated with plant species that can tolerate the changing environmental conditions, and use these markers in traditional breeding programs. The use of markers would result in a greater chance of incorporating desirable segments of DNA into crop species and with a greatly reduced lag period between the initial crossing and cultivar release than if only traditional breeding methodology was employed.

References

1. U.S. Congress Office of Technology Assessment. A New Technological Era for American Agriculture. OTA-F-474. Washington, DC: U.S. Government Printing Office, 1992.
2. Gasser CS, Fraley RT. Genetically engineering plants for crop improvement. Science 1989; 244:1293-1299.
3. Delannay X, Bauman TT, Beighley DH et al. Yield evaluation of a glyphosate-tolerant soybean line after treatment with glyphosate. Crop Sci 1995; 35:1461-1467.
4. Padgette SR, Kolacz KH, Delannay X et al. Development, identification and characterization of a glyphosate-tolerant soybean line. Crop Sci 1995; 35:1451-1461.
5. Nelson RS, McCormick SM, Delannay X et al. Virus tolerance, plant growth and field performance of transgenic tomato plants expressing coat protein from tobacco mosaic virus. BioTechnology 1988; 6:403-409.
6. Broglie K, Chet I, Holliday M et al. Transgenic plant with enhanced resistance to the fungal pathogen *Rhizoctonia solani*. Science 1991; 254:1194-1197.
7. Bayley C, Trolinder N, Ray C et al. Engineering 2,4-D resistance into cotton. Theor Appl Genet 1992; 83:645-649.
8. Feitelson JS, Payne J, Kim L. *Bacillus thuringensis*: Insects and beyond. BioTechnology 1992; 10:271-276.
9. Gray J, Picton S, Shabbeer J et al. Molecular biology of fruit ripening and its manipulation with antisense genes. Plant Mol Biol 1992; 19:69-87.
10. Caldwell MM, Teramura AH, Tevini M. The changing solar ultraviolet climate and the ecological consequences for higher plants. Trends Ecol Evolut 1989; 4:363-367.
11. Krupa SV, Kickert RN. The Effects of Elevated Ultraviolet (UV)-B Radiation on Agricultural Production. A peer reviewed critical assessment report submitted to the Formal Commission on "Protecting the Earth's Atmosphere" of the German Parliament, Bonn, Federal Republic of Germany, 1993.

12. Teramura AH, Sullivan JH. Potential impacts of increased solar UV-B on global plant productivity. In: Riklis E, ed. Photobiology: The Science and Its Applications. New York: Plenum Press, 1991:625-634.
13. Limpert E, Andrivon D, Fishbeck G. Virulence patterns in populations of *Erysiphe graminis* f. sp. *hordei* in Europe in 1986. Plant Pathol 1990; 39:402-415.
14. Runeckles VC, Krupa SV. The impact of UV-B and ozone on terrestrial vegetation. Environ Pollut 1994; 83:191-213.
15. Strid Å, Soon Chow W, Anderson JM. UV-B damage and protection at the molecular level in plants. Photosyn Res 1994; 39:475-489.
16. Ellison MJ, Childs JD. Pyrimidine dimers induced in *Escherichia coli* DNA by ultraviolet radiation present in sunlight. Photochem Photobiol 1981; 50:69-73.
17. Sutherland BM, Takayanagi S, Sullivan JH et al. Plant responses to changing environmental stress: Cyclobutyl pyrimidine dimer repair in soybean leaves. Photochem Photobiol 1996; 64(3):464-468.
18. Ballaré CL, Scopel AL, Stapleton AE et al. Solar ultraviolet-B radiation affects seedling emergence, DNA integrity, plant morphology, growth rate and attractiveness to herbivore insects in *Datura ferox*. Plant Physiol 1996; 112:161-170.
19. Miguel AG, Tyrrell RM. Induction of oxygen-dependant lethal damage by monochromatic UV-B (313 nm) radiation: Strand breakage, repair and cell death. Carcinogenesis 1983; 4:375-380.
20. Peak JG, Peak MJ, Sikorski RS et al. Induction of DNA-protein crosslinks in human cells by ultraviolet and visible radiations: Action spectrum. Photochem Photobiol 1985; 41:295-302.
21. Childs JD, Paterson MC, Smith BP et al. Evidence for a near UV-induced photoproduct of 5-hydroxymethylcytosine in bacteriophage T4 that can be recognized by endonuclease V. Mol Gen Genet 1978; 167:105-112.
22. Sancar A, Sancar GB. DNA repair enzymes. Annu Rev Biochem 1988; 57:29-67.
23. Bornman JF. Target sites of UV-B radiation in photosynthesis of higher plants. Photochem Photobiol 1989; 4:145-158.
24. Jordan BR, He J, Chow WS et al. Changes in mRNA levels and polypeptide subunits of ribulose 1,5-bisphosphate carboxylase in response to supplementary ultraviolet-B radiation. Plant, Cell Environ 1992; 15:91-98.
25. Tyrrell RM, Souza-Neto A. Lethal effects of natural solar-ultraviolet radiation in repair proficient and repair deficient strains of *Escherichia coli*: Actions and interactions. Photochem Photobiol 1981; 34:331-338.
26. Tyrrell RM. Cell inactivation and mutagenesis by solar ultraviolet radiation. In: Helene C, Charlier M, Montenay-Garestier T et al, eds. Trends in Photobiology. New York: Plenum Press, 1982.
27. Webb RB, Tuveson RW. Differential sensitivity to inactivation of NUR and NUR[+] strains of *Escherichia coli* at six selected wavelengths

in the UV-A, UV-B, and UV-C ranges. Photochem Photobiol 1982; 36:525-530.
28. Okaichi K, Kajitani N, Nakajima K et al. DNA damage and its repair in *Dictyostelium discoideum* irradiated by health lamp light (UV-B). Photochem Photobiol 1989; 50:69-73.
29. Keszenman-Pereyra D. Repair of UV-damaged incoming plasmid DNA in *Saccharomyces cerevisiae*. Photochem Photobiol 1990; 51:331-342.
30. Jordan BR, James PE, Strid Å et al. The effect of ultraviolet-B radiation on gene expression and pigment composition in etiolated and green pea leaf tissue: UV-B-induced changes are gene-specific and dependent upon the development stage. Plant, Cell Environ 1994; 17:45-54.
31. Herrera-Estrella L, Simpson J. Foreign gene expression in plants. In: Shaw CH, ed. Plant Molecular Biology—A Practical Approach. Washington, DC: IRL Press, 1988:131-160.
32. Simpson J, Van Montagu M, Herrera-Estrella L. Photosynthesis-associated gene families: Differences in response to tissue-specific and environmental factors. Science 1986; 233:34.

Appendix 1

Common and Scientific Names of Crop Plants in the FAO Database

Common name	Scientific name
Wheat	*Triticum aestivum, T. durum, T. spelta*
Rice	*Oryza sativa*
Barley	*Hordeum disticum, H. hexasticum, H. vulgare*
Maize	*Zea mays*
Rye	*Secale cereale*
Oats	*Avena sativa*
Millet	*Echinocloa frumentacea, Eleusine coracana, Eragrostis abyssinica, Panicum miliaceum, Paspalum scrobiculatum, Pennisetum glaucum, Setaria italica*
Sorghum	*Sorghum dura, S. guineense, S. vulgare*
Buckwheat	*Fagopyrum esculentum*
Triticale	*Triticum* spp. x *Secale* spp.
Quinoa	*Chenopodium quinoa*
Fonio	*Digitaria exilis, D. fonio, D. iburua*
Canary seed	*Phalaris canariensis*
Mixed grain	
Potatoes	*Solanum tuberosum*
Sweet potatoes	*Ipomoea batatas*
Cassava	*Manihot esculenta,* syn. *M. utilissima, M. palmata,* syn. *M. dulcis*
Yautia (coco yam)	*Xanthosoma* spp.; *malanga, new cocoyam, ocumo, sagittifolium*
Taro (coco yam)	*Colocasia esculenta*
Yams	*Dioscorea* spp.
Sugarcane	*Saccharum officinarum*
Sugar beets	*Beta vulgaris* var. *altissima*
Beans, dry	*Phaseolus aconitifolius, Ph. acutifolius, Ph. angularis, Ph. aureus, Ph. calcaratus, Ph. coccineus, Ph. lunatus, Ph. mungo, Ph. vulgaris*
Broad beans, dry	*Vicia faba* (var. *equina;* var. *major;* var. *minor*)

Elevated Ultraviolet (UV)-B Radiation and Agriculture, by Sagar V. Krupa, Ronald N. Kickert and Hans-Jürgen Jäger.
© 1998 Springer-Verlag and Landes Bioscience.

Common name	Scientific name
Peas, dry	*Pisum arvense, P. sativum*
Chickpeas	*Cicer arietinum*
Cow peas, dry	*Dolichos sinensis; Vigna sinensis*
Pigeon peas	*Cajanus cajan*
Lentils	*Ervum lens; Lens esculenta*
Bambara beans	*Voandzeia subterranea*
Vetches	*Vicia sativa*
Lupins	*Lupinus* spp.
Brazil nuts	*Bertholletia excelsa*
Cashew nuts	*Anacardium occidentale*
Chestnuts	*Castanea* spp.: *C. sativa; C. vesca; C. vulgaris*
Almonds	*Amygdalus communis; Prunus amygdalus; P. communis*
Walnuts	*Jugland regia*
Pistachios	*Pistacia vera*
Kolanuts	*Cola acuminata; C. nitida; C. vera*
Hazelnuts	*Corylus avellana*
Areca nuts (betel)	*Areca catechu*
Soybeans	*Glycine soja*
Groundnuts in shell	*Arachis hypogaea*
Coconuts	*Cocos nucifera*
Oil palm fruit	*Elaeis guineensis*
Palm kernels	*Elaeis guineensis; Orbignya speciosa*
Palm oil	*Elaeis guineensis*
Olives	*Olea europaea*
Karite nuts (sheanuts)	*Butyrospermum parkii*
Castor beans	*Ricinus communis*
Sunflower seeds	*Helianthus annuus*
Rapeseed	*Brassica napus* var. *oleifera*
Tung nuts	*Aleurites cordata; A. fordii*
Safflower seed	*Carthamus tinctorius*
Sesame seed	*Sesamum indicum*
Melonseed	*Cucumis melo*
Mustard seed	*Brassica alba; B. hirta; B. nigra; Sinapis alba; S. nigra*
Poppy seed	*Papaver somniferum*
Tallowtree seeds	*Sapium sebiferum; Shorea aptera; S. stenocarpa; Stillingia sebifera*
Vegetable tallow	*Sapium sebiferum; Shorea aptera; S. stenocarpa; Stillingia sebifera*
Stillingia oil	*Stillingia sebifera*
Kapok fruit	*Ceiba pentandra*
Kapokseed in shell	*Ceiba pentandra*
Seed cotton	*Gossypium* spp.
Cottonseed	*Gossypium* spp.
Linseed	*Linum usitatissimum*
Hempseed	*Cannabis sativa*
Cabbages	*Brassica chinensis; B. oleracea* all var. except *botrytis*
Artichokes	*Cynara scolymus*
Asparagus	*Asparagus officinalis*

Appendix 1

Common name	Scientific name
Lettuce	*Cichorium endivia* var. *crispa*; *C. endivia* var. *latifolia*; *C. intybus* var. *foliosum*; *Lactuca sativa*
Spinach	*Atriplex hortensis; Spinacia oleracea; Tetragonia espansa*
Tomatoes	*Lycopersicon esculentum*
Cauliflower	*Brassica oleracea* var. *botrytis*, subvariety *cauliflora* and *cymosa*
Pumpkins; squash; gourds	*Cucurbita* spp.
Cucumbers and gherkins	*Cucumis sativus*
Eggplants	*Solanum melongena*
Chillies+peppers; green	*Capsicum annuum; C. fructescens; Pimenta officinalis*
Onions+shallots; green	*Allium ascalonicum; A. cepa; A. fistulosum*
Onions, dry	*Allium cepa*
Garlic	*Allium sativum*
Leeks & other alliac. veg.	*Allium porrum; A. schoenoprasum*
Beans, green	*Phaseolus* and *Vigna* spp.
Peas, green	*Pisum sativum*
Broad beans, green	*Vicia faba*
String beans	*Phaseolus vulgaris*
Carrots	*Daucus carota*
Okra	*Abelmoschus esculentus*
Green corn	*Zea mays* var. *saccharata*
Mushrooms	*Agaricus truffles campestris; Boletus edulis; Morchella* spp.; *Tuber magnatum*
Chicory roots	*Cichorium intybus; C. sativum*
Carobs	*Ceratonia siliqua*
Bananas	*Musa cavendishii; M. nana; M. sapientum*
Plantains	*Musa paradisiaca*
Oranges	*Citrus aurantium; C. sinensis*
Tang., mand., clement., satsma	*Citrus reticulata; C. unshiu*
Lemons & limes	*Citrus aurantifolia; C. limetta; C. limon*
Grapefruit & pomelo	*Citrus grandis; C. maxima; C. paradisi*
Apples	*Malus communis; M. pumila; M. sylvestris; Pyrus malus*
Pears	*Pyrus communis*
Quinces	*Cydonia oblonga; C. vulgaris; C. japonica*
Apricots	*Prunus armeniaca*
Sour cherries	*Cerasus acida; Prunus cerasus*
Cherries	*Cerasus avium; Prunus avium*
Peaches & nectarines	*Amygdalus persica; Persica laevis; Prunus persica*
Plums	*Prunus domestica; P. spinosa*
Strawberries	*Fragaria* spp.
Raspberries	*Rubus idaeus*
Gooseberries	*Ribes grossularia*
Currants	*Ribes nigrum; R. rubrum*
Blueberries	*Vaccinium corymbosum; V. myrtillus*
Cranberries	*Vaccinium macrocarpon; V. oxycoccus*
Grapes	*Vitis vinifera*

Common name	Scientific name
Watermelons	*Citrullus vulgaris*
Cantaloupes+other melons	*Cucumis melo*
Figs	*Ficus carica*
Mangoes	*Mangifera indica*
Avocados	*Persea americana*
Pineapples	*Ananas comosus; A. sativ*
Dates	*Phoenix dactylifera*
Persimmons	*Diospyros kaki; D. virginiana*
Cashewapple	*Anacardium occidentale*
Kiwi fruit	*Actinidia chinensis*
Papayas	*Carica papaya*
Straw; husks	
Maize for forage+silage	*Zea mays*
Sorghum for forage+silage	*Sorghum* spp.
Ryegrass; forage+silage	*Lolium* spp.
Alfalfa for forage+silage	*Medicago* spp.
Cabbage for fodder	*Brassica* spp.
Pumpkins for fodder	*Cucurbita* spp.
Beets for fodder	*Beta* spp.
Carrots for fodder	*Daucus carota*
Swedes for fodder	*Brassica* spp.
Clover for forage+silage	*Trifolium* spp.
Vegetables+roots; fodder	
Coffee, green	*Coffea* spp. *(arabica, robusta, liberica)*
Cocoa beans	*Theobroma cacao*
Tea	*Camellia sinensis; Thea assaamica; T. sinensis*
Mate	*Ilex paraguayensis*
Hops	*Humulus lupulus*
Pepper; white/long/black	*Piper longum; P. nigrum*
Pimento+allspice	*Capsicum annuum; C. frutescens; Pimenta officinalis*
Vanilla	*Vanilla planifolia; V. pompona*
Cinnamon (canella)	*Cinnamomum cassia; C. zeylanicum*
Cloves; whole+stems	*Caryophyllus aromaticus; Eugenia caryophyllata*
Nutmeg; mace; cardamons	*Aframomum angustifolium; A. hambury; A. melegueta; Amomun aromaticum; A. cardamomum; Elettaria cardamomum; Myristica fragrans*
Anise; badian; fennel	*Carum carvi; Coriandrum sativum; Cuminum cyminum; Foeniculum vulgare; Illicium verum; Juniperus communis; Pimpinella anisum*
Ginger	*Zingiber officinale*
Peppermint	*Mentha* spp.: *M. piperita*
Pyrethrum; dried flowers	*Chrysanthemum cinerariifolium*
Cotton lint	*Gossypium* spp.
Flax fibre & tow	*Linum usitatissimum*
Hemp fibre & tow	*Cannabis sativa*
Kapok fibre	*Ceiba pentandra*
Jute	*Corchorus capsularis; C. olitorius*

Appendix 1

Common name	Scientific name
Jute-like fibres	*Abroma augusta; Abutilon avicennae; Crotalaria juncea; Hibiscus cannabinus; H. sabdariffa; Urena lobata; U. sinuata*
Ramie	*Boehmeria nivea; B. tenacissima*
Sisal	*Agave sisalana*
Abaca (manila hemp)	*Musa textilis*
Coir	*Cocos nucifera*
Tobacco leaves	*Nicotiana tabacum*
Natural rubber	*Hevea brasiliensis*
Natural gums	*Achras zapota; Dieva costulana; Manihot glaziovii; Manilkara bidentata; Palachium gutta; Parthenium argentatum*
Hay non-leguminous	
Hay (clover; lucerne; etc.)	*Trifloium* spp.; *Medicago* spp.
Hay (unspecified)	
Range pasture	
Improved pasture	

APPENDIX 2

Countries in the FAO Database

Countries of Latin America and the Caribbean

Anguilla	Guyana
Antigua and Barbuda	Haiti
Argentina	Honduras
Aruba	Jamaica
Bahamas	Martinique
Barbados	Mexico
Belize	Montserrat
Bolivia	Netherlands Antilles
British Virgin Islands	Nicaragua
Brazil	Panama
Cayman Islands	Paraguay
Chile	Peru
Colombia	Puerto Rico
Costa Rica	Saint Lucia
Cuba	South Georgia
Dominica	St Kitts-Nevis
Dominican Republic	St Vincent
Ecuador	Suriname
El Salvador	Trinidad and Tobago
Falkland Islands	Turks & Caicos Islands
French Guiana	U.S. Virgin Islands
Grenada	Uruguay
Guadalupe	Venezuela
Guatemala	

Developing countries of Africa

Algeria	Burundi
Angola	Cameroon
Benin	Cape Verde
Botswana	Central African Republic
British Indian Ocean Territory	Chad
Burkina Faso	Comoros

Elevated Ultraviolet (UV)-B Radiation and Agriculture, by Sagar V. Krupa, Ronald N. Kickert and Hans-Jürgen Jäger.
© 1998 Springer-Verlag and Landes Bioscience.

Congo
Cote d'Ivoire
Djibouti
Egypt
Equatorial Guinea
Eritrea
Ethiopia
Gabon
Gambia
Ghana
Guinea
Guinea-Bissau
Kenya
Lesotho
Liberia
Libya
Madagascar
Malawi
Mali
Mauritania
Mauritius
Mayotte
Morocco

Mozambique
Namibia
Niger
Nigeria
Reunion
Rwanda
São Tomé and Príncipe
Senegal
Seychelles
Sierra Leone
Somalia
St Helena
Sudan
Swaziland
Tanzania
Togo
Tunisia
Uganda
Western Sahara
Zaire
Zambia
Zimbabwe

Countries of Western Europe

Andorra
Austria
Bel-lux
Denmark
Faeroe Islands
Finland
France
Germany
Germany Fr
Germany Nl
Gibraltar
Greece
Holy See
Iceland
Ireland

Italy
Liechtenstein
Luxembourg
Malta
Monaco
Netherlands
Norway
Portugal
San Marino
Spain
Svalbard Etc
Sweden
Switzerland
United Kingdom

Countries of the transition markets (including Eastern Europe)

Albania
Armenia
Azerbaijan
Belarus
Bosnia Herzegovina
Bulgaria
Croatia
Czech Republic
Estonia
Georgia
Hungary
Kazakhstan
Kyrgyzstan
Latvia

Lithuania
Macedonia
Moldova Republic
Poland
Romania
Russian Federation
Slovakia
Slovenia
Tajikistan
Turkmenistan
Ukraine
Uzbekistan
Yugoslav Fr

Developing countries of Asia

Afghanistan
Bahrain
Bangladesh
Bhutan
Brunei Darsm
Cambodia
China Main
China Taiwan
Cyprus
East Timor
Gaza Strip
Hong Kong
India
Indonesia
Iran
Iraq
Jordan
Korea Democratic Peoples Republic
Korea Republic
Kuwait
Laos

Lebanon
Macau
Malaysia
Maldives
Mongolia
Myanmar
Nepal
Oman
Pakistan
Philippines
Qatar
Saudi Arabia
Singapore
Sri Lanka
Syria
Thailand
Turkey
United Arab Emirates
Viet Nam
West Bank
Yemen

Countries of Oceania

American Samoa	New Zealand
Australia	New Caledonia
Canton Island	Niue
Christmas Island	Norfolk Island
Cocos Island	Pacific Island
Cook Island	Palau
Fiji	Papua New Guinea
French Polynesia	Pitcairn
Guam	Samoa
Johnston Island	Solomon Islands
Kiribati	Tokelau
Marshall Islands	Tonga
Micronesia	Tuvalu
Midway Island	Us Minor Island
North Marianas	Vanuatu
Nauru	Wake Island
	Wallis and Futuna

APPENDIX 3

Common and Scientific Names of Plant Species (excluding Chapter 8)

Common name	Scientific name
Alfalfa	*Medicago sativa* L.
Alpine (whiproot) clover	*Trifolium dasyphyllum* L.
Alyce clover	*Alysicarpus vaginalis* (L.) DC
Apple	*Malus pumila* Miller
Artichoke	*Cynara scolymus* L.
Asparagus	*Asparagus officinalis* L.
Aster	*Aster tripolium* L.
Banana	*Musa* spp. L.
Barley	*Hordeum vulgare* L.
Bean	*Phaseolus vulgaris* L.
Bermuda grass	*Cynodon dactylon* (L.) Pers.
Bilberry	*Vaccinium myrtillus* L.
Bluebell	*Hyacinthoides non-scripta* (L.) Rothm.
Blueberry	*Vaccinium* spp. L.
Broad bean	*Vicia faba* L.
Broccoli	*Brassica oleracea* L. var. *botrytis*
Brussels sprouts	*Brassica oleracea* L. var. *gemmifera*
Cabbage	*Brassica oleracea* L. var. *capitata*
Cantaloupe	*Cucumis melo* L.
Carnation	*Dianthus caryophyllus* L.
Carrot	*Daucus carota* L.
Cassava	*Manihot esculenta* Crantz
Cauliflower	*Brassica oleracea* L. var. *botrytis*
Celery	*Apium graveolens* L.
Chards	*Beta vulgaris* L. var. *cicla* (L.) Koch
Chrysanthemum	*Chrysanthemum morifolium* L.
Chufa	*Cyperus esculentus* L.
Cloudberry	*Rubus chamaemorus* L.
Clover	*Trifolium* spp. L., *Trifolium subterraneum* L.
Coconut	*Cocos nucifera* L.
Coffee	*Coffea* spp. L.
Coleus	*Solenostemon* spp. Thonn.

Elevated Ultraviolet (UV)-B Radiation and Agriculture, by Sagar V. Krupa, Ronald N. Kickert and Hans-Jürgen Jäger.
© 1998 Springer-Verlag and Landes Bioscience.

Common name	Scientific name
Collards	*Brassica oleracea* L. var. *acephala*
Common heather	*Calluna vulgaris* (L.) Hull
Corn	*Zea mays* L.
Cotton	*Gossypium hirsutum* L.
Cowpea	*Vigna sinensis* Savi, *V. unguiculata* L.
Cress	*Lepidium sativum* L.
Cucumber	*Cucumis sativus* L.
Cyclamen	*Cyclamen* spp. L.
Digit grass	*Digitaria decumbens* Stent
Eggplant	*Solanum melongena* L.
Endive	*Cichorium endivia* L.
English oak	*Quercus robur* L.
European beech	*Fagus sylvatica* L.
Floribunda rose	*Rosa* spp. L.
Grape	*Vitis vinifera* L.
Groundnuts	*Arachis hypogaea* L.
Hemp	*Cannabis sativa* L.
Ivy geranium	*Geranium* spp. L.
Kale	*Brassica oleracea* L. var. *acephala*
Kentucky bluegrass	*Poa pratensis* L.
Kohlrabi	*Brassica oleracea* L. var. *gongylodes*
Lemon, rough	*Citrus jambhiri* L.
Lettuce	*Lactuca sativa* L.
Loblolly pine	*Pinus taeda* L.
Marigold	*Calendula officinalis* L.
Melon	*Cucumis melo* L.
Millet	*Setaria italica* (L.) Pal
Muskmelon	*Cucumis melo* L.
Mustard	*Brassica* spp. L.
Nasturtium	*Tropaeolum* spp. L.
Nectarine	*Prunus persica* L. var. *nectarina* (Aiton) Maxim.
Oats	*Avena sativa* L.
Oil palm	*Elaeis guineensis* Jacq.
Okra	*Abelmoschus esculentus* L., *Hibiscus esculentus* L.
Olive	*Olea europaea* L.
Onion	*Allium cepa* L.
Orange	*Citrus sinensis* L.
Orchard grass	*Dactylis glomerata* L.
Papaya	*Carica papaya* L.
Parsnip	*Pastinaca sativa* L.
Patience dock	*Rumex patientia* L.
Pea	*Pisum sativum* L.
Peaches	*Prunus persica* (L.) Batsch
Peanut	*Arachis hypogaea* L.
Pepper	*Capsicum frutescens* L.
Petunia	*Petunia hybrida* Vilm.

Common name	Scientific name
Plantains	*Plantago* spp. L.
Poinsettia	*Euphorbia pulcherrima* Willd. ex Klotzsch
Potato	*Solanum tuberosum* L.
Pumpkin	*Cucurbita pepo* L.
Radish	*Raphanus sativus* L.
Rapeseed	*Brassica napus* L.
Red clover	*Trifolium pratense* L.
Red raspberry	*Rubus strigosus* L.
Rhubarb	*Rheum rhaponticum* L.
Rice	*Oryza sativa* L.
Richardson geranium	*Geranium richardsonii* L.
Rose	*Rosa* spp. L.
Rutabaga	*Brassica napus* L. var. *napobrassica*
Rye	*Secale cereale* L.
Ryegrass	*Lolium perenne* L.
Salt marsh grass	*Elymus athericus* L.
Snapbean	*Phaseolus vulgaris* L.
Snapdragon	*Antirrhinum majus* L.
Sorghum	*Sorghum vulgare* (L.) Moench
Soybean	*Glycine max* (L.) Merr.
Spinach	*Spinacia oleracea* L.
Squash	*Cucurbita* spp. L.
Strawberry	*Fragaria* spp. L.
Sugar beet	*Beta vulgaris* L.
Sugarcane	*Saccharum officinarum* L.
Sunflower	*Helianthus annuus* L.
Sweet corn	*Zea mays* L. var. *saccharata*
Sweet gum	*Liquidambar styraciflua* L.
Sweet pepper	*Capsicum frutescens* L.
Sweet potato	*Ipomoea batatas* (L.) Lam.
Swiss chards	*Beta vulgaris* subsp. *cicla* (L.) Koch
Taro	*Colocasia esculenta* (L.) Schott
Tobacco	*Nicotiana tabacum* L.
Tomato	*Lycopersicon esculentum* Miller
Triticale	*Triticum* spp. L. x *Secale* spp. L.
Turnip	*Brassica rapa* L. var. *rapifera*
Watermelon	*Citrullus vulgaris* L.
Wheat	*Triticum aestivum* L.
White mustard	*Sinapis alba* L.
Wild oat grass	*Avena fatua* L.
Willow	*Salix dasyclados* L.
Wood anemone	*Anemone nemorosa* L.
Yams	*Dioscorea* spp. L.
Yellow alyssum	*Alyssum alyssoides* (L.) L.

Index

A

Abelmoschus, 203, 275, 284
Abutilon, 203, 277
Action spectra, 35-37, 42, 43, 52, 53, 55, 67, 84, 85, 87, 240
Adaptation
  eco-physiological, 141
Additive main effect and multiplicative (AMMI), 195
Aegilops, 162
Africa, 24, 152-157, 162, 163, 189, 197, 221, 225, 226, 228, 230, 232, 234-236, 238, 279
Agricultural production
  current, 224
  world, 219, 224
Agropyron, 158, 162
Allium, 119, 275, 284
Amaranthus, 151, 153-156, 160, 203, 204
Apium, 283
Arachis, 109, 274, 284
Asia, 39, 56, 59, 153, 154, 156, 157, 160-163, 189, 192, 206, 221, 225, 226, 228, 230, 232, 234-236, 238, 281
Asparagus, 117, 119, 121, 229, 238, 250, 274, 283
Aster, 151, 190, 193, 201, 283
Avena, 80, 108, 109, 151, 153, 158, 161, 189, 211, 267, 273, 284, 285

B

Beetles, 175, 210
Beta, 119, 152, 168, 209, 225, 273, 276, 283, 285
Biomass accumulation, 108-110, 119, 120, 188, 203, 204, 235
Biotroph
  facultative, 168
Brassica, 109, 119, 154, 155, 157, 158, 225, 274-276, 283-285
Brewer spectrophotometer, 70

C

Capsella, 152, 157, 162
Capsicum, 118, 275, 276, 284, 285
Caribbean, 153, 157, 160, 163, 203, 225, 226, 228, 230, 232, 234-236, 238, 279
Carica, 153, 201, 276, 284
Cercospora, 168, 209
Cercospora leaf spot, 168
Chenopodium, 151, 154-156, 160, 161, 273
Chromophore, 139, 141
Cinnamic acid, 139
Citrus, 118, 153, 163, 170, 175, 201, 203, 230, 275, 284
Cladosporium, 168, 169, 174
Climate database, 2
Clouds, 11, 38, 39, 44, 50, 51, 53, 54, 56, 57, 134
CO_2, 135, 152, 157, 160, 161, 163, 171-173, 181, 183, 184, 186, 189, 190, 192, 196, 201-202
Coffea, 153, 276, 283
Colletotrichum, 168-170, 174
Competition
  density effect, 149
  inter-species, 147, 149, 151
  intra-species, 109, 148, 222
  plant, 150, 157
  plant species, 154
  UV-B effect, 93, 95, 109, 135, 136, 172, 195, 200, 223
Crop-weed competition, 149
Cropping systems, 248, 252, 257-259, 264
  genetic engineering, 250, 256, 263-265, 269
  losers, 153, 160, 163
  loss, 11, 12, 14, 147, 149, 151, 248, 258, 259, 263, 265, 269
  primary effects, UV-B, 106
  weeds, 147-151
  winners, 153, 163
Cucumis, 90, 118, 168, 169, 189, 274-276, 283, 284
Cucurbita, 118, 275, 276, 285
Cyclobutane, 86, 139, 264, 266
Cynara, 119, 274, 283
Cynodon, 119, 153, 155-157, 161, 163, 283
Cyperus, 119, 156, 158, 161, 283

D

Dactylis, 119, 284
Dictyostelium, 265
Digitaria, 119, 156, 157, 162, 273, 284
Dimorphotheca, 201
Diplocarpon, 169, 174
Disease
    incidence, 171, 172
    multiple stress, 182, 186, 201, 207, 208
DNA
    damage, 47, 67, 86, 88, 265, 266
    damage, repair genes of, 265, 267
    transfer, Agrobacterium, 266
    transfer, direct, 266
Dobson units, 11
Dose
    biologically effective, 84
    effective, 37, 89
    exposure, 67, 89, 190, 198
Dosimetry
    CO_2, 89
    O_3, 89
    UV-B, 75, 89

E

Eastern Europe, 152
Echinochloa, 152, 155, 157, 158, 160-162
Ecological genetics, 247
Elaeis, 153, 274, 284
Elite germplasm, 250, 252, 253, 255, 257, 258, 260
Elymus, 162, 192, 285
Epidermal transmittance, 150
Epilachna, 176
Erythema, 20, 21, 23, 24, 28, 29, 32, 34, 35, 38, 42-44, 51, 56-58, 67-69, 82, 86
Escherichia, 86, 267
Europe, eastern, 153, 163, 189, 225, 235, 281
Excision, 265, 266
Expectations
    long-term, 258
    short-term, 259
Exposure dynamics, 89, 90, 93
Exposure kinetics, O_3, 89

F

Ficus, 201, 276
Filter, UV, 29, 67, 68, 70, 73-76, 78, 80, 82, 83, 107, 134, 137
Flavonoids, 108, 137, 142, 172
Free radicals, 139
Fusarium, 170, 171, 174

G

Galinsoga, 155, 156, 160
Gene expression, 268
General circulation model (GCM), 56, 221
Genetic diversity, 248-251, 253, 260, 268
Genetic markers, 268-270
Genetic uniformity, 248, 250
Geum, 153
Global climate, 1, 134, 167, 258
Global radiation, 4, 59
Glycine, 76, 80, 86, 107, 175, 189, 235, 249, 263, 274, 285
Gossypium, 90, 153, 201, 206, 225, 274, 276, 284
Greenhouse effect, 4

H

Hardy-Weinberg equilibrium, 250, 251
Helianthus, 109, 173, 196, 274, 285
Hibiscus, 118, 154, 277, 284
Hordeum, 14, 108, 153, 240, 260, 263, 273, 283

I

Indole acetic acid, 137
Insect
    herbivory, UV-B effects on, 176, 210
    multiple stress, 210
Interactions
    CO_2-light-temperature, 203
    CO_2-nutrient, 205
    CO_2-O_3, 204
    crop-weed, 147
    environment-crop, 181
    types of, 182, 185
    UV-B-CO_2, 190
    UV-B-moisture, 200
    UV-B-nutrient, 201
    UV-B-O_3, 196
    UV-B-visible light, 198

Index

Interactive plant response, 185
International Futures Simulation (IFS), 240, 243
Interspecific uniformity, 259
Intraspecific uniformity, 260
Irradiance
  effective, 31, 36, 44, 57-59, 85
  erythemally weighted, 44

J

Junonia, 175

L

Lactuca, 119, 189, 201, 275, 284
Lamps, UV, 73, 75
Latin America, 153, 160, 163, 203, 225, 235, 236, 238, 279
Lolium, 140, 276, 285
Lycopersicon, 118, 161, 176, 189, 190, 193, 275, 285

M

Magnaporthe, 169, 174
Malus, 222, 275, 283
Mating, random, 250
Mechanism of action, 168
Medicago, 90, 119, 153, 266, 276, 277, 283
Model calculations, 15, 33, 45
Moisture-temperature interaction, 207
Multi-point model, 91
Mutation, 170

N

Nicotiana, 119, 140, 161, 194, 196, 225, 277, 285
NO_2, photolysis of, 133

O

Oceania, 154-157, 221, 225, 226, 228, 230, 232, 234-236, 238, 240, 282
Olea, 201, 225, 274, 284
Oryza, 78, 109, 161, 189, 225, 257, 267, 273, 285
Ozone column, 160-161

P

Pastinaca, 119, 210, 284
Pathogen, incidence, 168, 171
Penicillum, 170
Pest incidence, 167
Petunia, 116, 120, 193, 196, 284
Phaseolus, 79, 109, 139, 173, 189, 194, 197, 198, 220, 273, 275, 283, 285
Phenylpropanoids, 137
Photochemical model, 15
Photoreactivation, 266
Photorepair, 73, 87, 107, 139, 142, 189
Photosynthetically active radiation (PAR), 73, 107, 148, 185
Photosystem II, 85
Pigment accumulation, 137, 176
Pisum, 139
Plant breeding, 252, 254, 260, 268
Plant damage, 34, 47, 67, 69, 73, 88, 240
Plant effective UV-B, 35, 46, 52, 58, 59, 149, 150, 152, 153, 160-163, 197, 198, 200
Plant response to UV-B, 90, 162, 189
Plant transformation, 266
Plantago, 175, 285
Poa, 119, 140, 153, 155, 284
Polar vortex, arctic, 13, 18
Polygonum, 155, 156, 158, 161
Polymorphism, fine scale, 250
Population
  crop, 149, 247, 250-254, 257
  genetics, 247, 250, 251
  plant, 247, 257
Populus, 141
Portulaca, 155, 156, 157 161-163
Prediction models, 211
Pseudomonas, 140
Puccinia, 173, 174, 209

Q

Quantitative genetic, 247, 251

R

Radiation amplification factor (RAF), 12, 14, 39, 46, 47, 85, 86
Radiative transfer calculation, 18
Raphanus, 90, 107, 140, 141, 158, 200, 285
Rheum, 119, 285
Ribulose-1,5-bisphosphate carboxylase, 264
Rice blast, 169, 170
Robertson-Berger, 33, 42, 43, 58, 67-69, 87, 88, 134
Rubus, 81, 118, 275, 283, 285
Rumex, 86, 88, 152, 154, 157, 158, 162, 284

S

Saccharomyces, 265
Saccharum, 119, 153, 220, 273, 285
Salix, 210, 285
Secale, 109, 153, 202, 273, 285
Sensitivity
 differential, 108
 ranking, 108, 235
Setaria, 109, 154, 156, 273, 284
Single-point model, 91
Solanum, 107, 118, 152, 155, 161, 189, 225, 273, 275, 284, 285
Solar radiation, 150, 211
Sorghum, 86, 90, 109, 111, 120, 153, 155-157, 159, 187, 189, 226, 232, 236, 250, 273, 276, 285
Spartina, 151
Spectroradiometer, 68, 70-72, 74
Spinacia, 119, 171, 275, 285
Square wave, 151
Stellaria, 152, 154, 155, 158, 160, 161, 163
Stratospheric ozone, 151, 168, 184, 192, 197, 198, 200, 209
Superoxide anion, 140
Superoxide dismutase (SOD), 140
Synergistic change, 182, 209
Synergistic effect, 186
Synergistic interactions, 182, 190, 207

T

t-DNA, 267
T-region, 267
TableCurve 3D, 186
Threshold-21 model, 242
Ti plasmid, 267
Total Ozone Monitoring Spectrometer (TOMS), 12
Transition market, 152, 153, 160, 162, 281
Trialeurodes, 176
Trichoplusia, 175
Trifolium, 119, 141, 205, 276, 283, 285
Triticum, 14, 80, 107, 151, 168, 186, 225, 253, 257, 267, 273, 285
Tropospheric vacuum cleaner, 133

U

Uncinula, 171, 174
Uromyces, 173, 174
UV-B damage, target of, 264
UV-B exposure, 33, 34, 38, 44-46, 56, 57, 80, 85, 106, 107, 137, 148, 149, 151, 163, 168, 170, 175, 186, 189, 195, 200, 235, 264
UV-B flux, 11, 18, 27, 36, 106-108, 133, 135, 141, 147
UV-B measurements, 42, 58, 59, 70
UV-B spatial variation, 50, 219

V

Vaccinium, 81, 118, 275, 283
Vigna, 109, 200, 201, 274, 275, 284
Vitis, 153, 171, 225, 275, 284

W

Weighting function, 73, 84, 85, 90
Western Europe, 2, 152, 153, 160, 163, 225, 235, 280
World agricultural price (WAP), 242
World agricultural production (WAPRO), 219, 224
World food stock (WSTK), 241, 242
World radiation network, 4

Z

Zea, 109, 162, 206, 225, 248, 267, 273, 275, 276, 284, 285

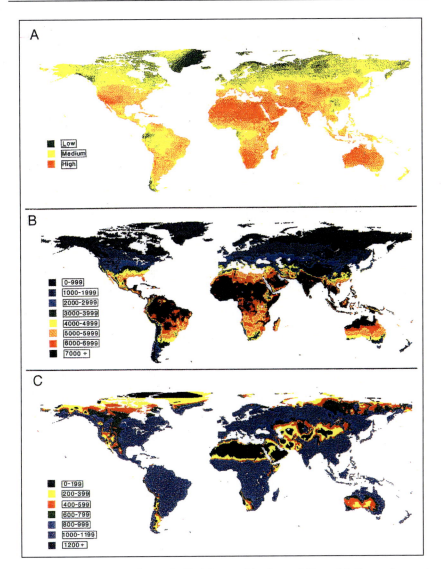

Fig. 1.1. (A) Average annual recorded bright sunshine hours.[3] Note: this figure does not include the level of total radiation or UV-B radiation. (B) Average annual growing degree days (above 5°C).[3] (C) Average annual precipitation (mm).[3]

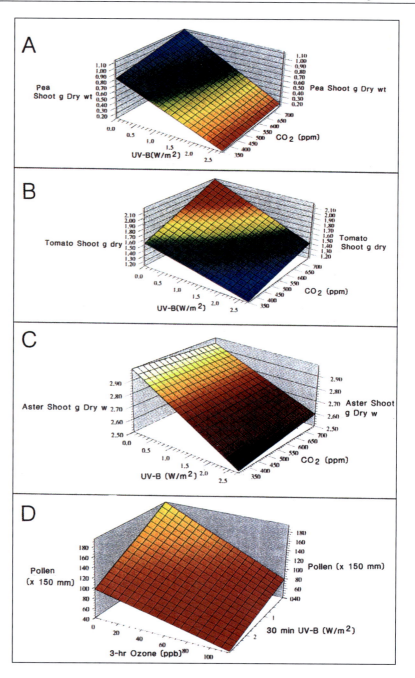

Fig. 7.1. (A) Relationship between shoot dry weight, UV-B and CO_2 exposures in pea (*Pisum sativum* L.).[24] (B) Relationship between shoot dry weight, UV-B and CO_2 exposures in tomato (*Lycopersicon esculentum* Miller).[24] (C) Relationship between shoot dry weight, UV-B and CO_2 exposures in aster (*Aster tripolium* L.).[24] (D) Relationship between pollen tube length, UV-B and O_3 exposures in petunia (*Petunia hybrida* Vilm.).[21]

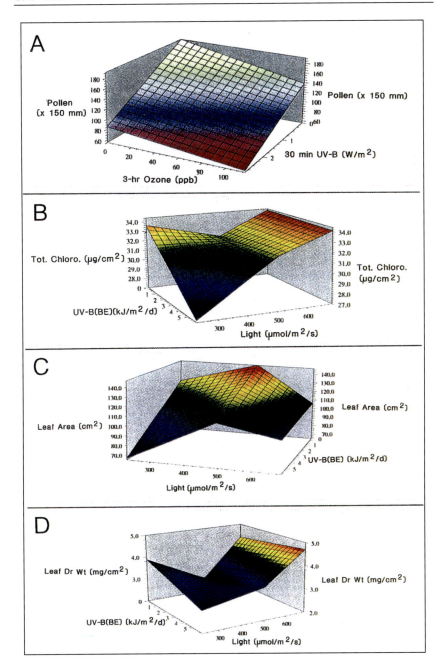

Fig. 7.2. (A) Relationship between pollen tube length, UV-B and O_3 exposures in tobacco (*Nicotiana tabacum* L.).[21] (B) Relationship between leaf total chlorophyll, UV-B and light exposures in bean (*Phaseolus vulgaris* L.).[36] (C) Relationship between leaf area, UV-B and light exposures in bean.[36] (D) Relationship between leaf dry weight, UV-B and light exposures in bean.[36]

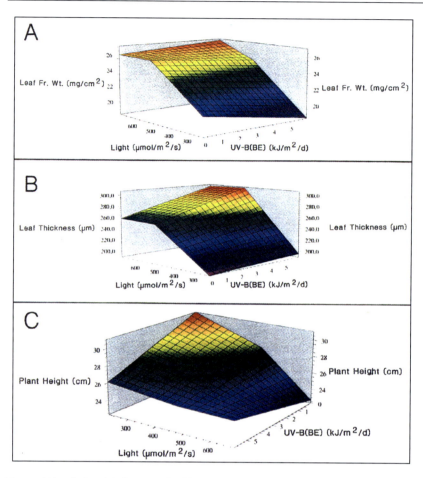

Fig. 7.3. (A) Relationship between leaf fresh weight, UV-B and light exposures in bean.[36] (B) Relationship between leaf thickness, UV-B and light exposures in bean.[36] (C) Relationship between plant height, UV-B and light exposures in bean.[36]

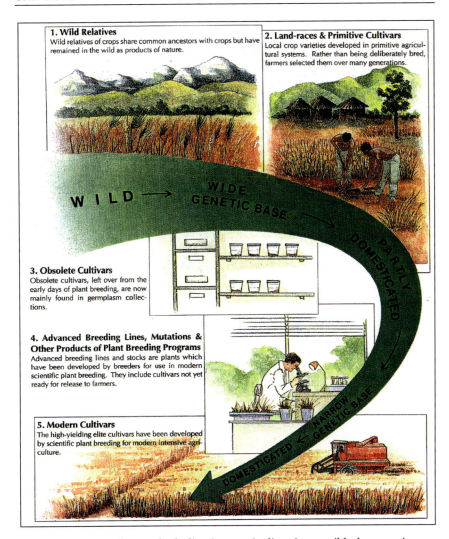

Fig. 9.1. Illustrating the steady decline in genetic diversity as wild plant species are increasingly domesticated.[2]